Quantum Field Theory with Application to Quantum Nonlinear Optics

Anatoliy K Prykarpatsky

University of Mining and Metallurgy of Krakow, Poland and
Ukrainian National Academy of Sciences, Ukraine

Ufuk Taneri

Eastern Mediterranean University, Turkey

Nickolai N Bogolubov, Jr

Steklov Mathematical Institute of
Russian Academy of Sciences, Russia

Quantum Field Theory with Application to Quantum Nonlinear Optics

World Scientific
New Jersey • London • Singapore • Hong Kong

Published by

World Scientific Publishing Co. Pte. Ltd.
5 Toh Tuck Link, Singapore 596224
USA office: Suite 202, 1060 Main Street, River Edge, NJ 07661
UK office: 57 Shelton Street, Covent Garden, London WC2H 9HE

British Library Cataloguing-in-Publication Data
A catalogue record for this book is available from the British Library.

QUANTUM FIELD THEORY WITH APPLICATION TO QUANTUM NONLINEAR OPTICS

ISBN 981-238-163-5
ISBN 981-238-164-3 (pbk)

Printed by FuIsland Offset Printing (S) Pte Ltd, Singapore

Preface

Many modern electronic devices are operating with optical signals generated due to the interaction of a working substance like crystal with an external laser radiation. As in most of these devices the interaction is nonlinear and the processes making them stable are quantum, it is important to device and use suitable approaches to studying their equilibrium radiative states subject to the external conditions imposed on a system. A great deal of studies were done concerning the problems of stability, bifurcation behaviour and dynamical properties of atomic and molecular systems having applications in optical bi-stability of a laser cavity with a nonlinear crystal medium, in some microelectronics and other devices based on nonlinear interaction processes with radiation.

Multi photon excitation states of poly-atomic molecules undergoing a self-interaction via Kerr effect related processes, are of great interest today; sucessful study of these are analytical and by means of modern quantum field theoretical tools. These and related topics are the main aim of the presented lecture notes developing the modern quantum field theory methods for the analysis of radiative states in a nonlinear quantum-optical system having important applications in electronics.

These notes being prepared for graduate mathematical physics and physics students can also be of interest to mathematicians involved in applied physics problems, physicists and chemists studying phenomena related with modern quantum-optical devices.

The first twelve chapters constitute Part I: 'Methods of Studying Quantum Optical Phenomena', the next six Part II: 'Nonlinear Quantum Optics Models and Their Applications'. Appendices to Part I are given at end of the book.

Acknowledgements

Of the authors, A.P. is cordially thankful to his wife Natalia for her expert help in preparing the manuscript and to Prof. D.L. Blackmore for valuable discussions; U.T is grateful to her son Niyazi (17) for his endless encouragement for her work and her assistant Arif Akkelesh for typesetting the manuscript. The authors are also thankful to Prof. M. Halilsoy for interesting discussions of the preliminary version of the lecture notes.

Contents

PART 1

METHODS OF STUDYING QUANTUM OPTICAL PHENOMENA

Chapter 1

Introduction to Quantum Mechanics in Microworld

Assume we are given a classical dynamical system describing the motion of a particle of mass m under some external field interaction. In general, as we know from classical mechanics, the motion of this particle object in the space-time $\mathbb{R}^3 \times \mathbb{R}_t$ is governed by the Fermat least action principle:

$$\delta S[\mathbf{x}] = 0\,, \tag{1.1}$$

where, by definition

$$S[\mathbf{x}] := \int_{t_1}^{t_2} \mathcal{L}(\mathbf{x}, \dot{\mathbf{x}}; t)dt \tag{1.2}$$

is an action and the function $\mathcal{L}(\mathbf{x}, \dot{\mathbf{x}}; t)$ is called a Lagrangian. The condition (1.1) results in the following Eulerian equation:

$$\frac{\partial \mathcal{L}}{\partial \mathbf{x}} - \frac{\partial}{\partial t}(\partial \mathcal{L} \partial \dot{\mathbf{x}}) = 0\,. \tag{1.3}$$

Having solved (1.3) at some initial conditions $\mathbf{x}|_{t=t_0} = \mathbf{x}_0$, $\dot{\mathbf{x}}|_{t=t_0} = \mathbf{v}_0$, the full dynamics of the particle under regard can be retrieved. Unfortuanetly the principle (1.1) is good, as shown by experiments, only at large enough distances a particle propagates in the space-time, and at large enough masses of particles or objects governed by (1.3).The case of tiny enough particles' propagation in space-time, needs a significantly different approach to treat their dynamics; this is based on new probabilistic considerations. As is well known, this idea was introduced for the first time by Schrödinger and developed further by R. Feynman in 1948 (Nobel prize of 1964).

Namely, the fundamental idea of describing the motion in microworld by means of a probability amplitude appeared to be fundamental and closely related with virtual classical motions of a particlelike object under regard. Before taking this new quantum ingredient into consideration, let us recall that the probability notion describing a degree of possibility of an event under regard is defined by means of some additive function subject to so called independent sets-events: $0 \leq p(A_i) \leq 1$ for all $i \in \mathbb{Z}_+$, and

$$p\left(\bigcup_{i\in\mathbb{Z}_+} A_i\right) = \sum_{i\in\mathbb{Z}_+} p(A_i)\,, \tag{1.4}$$

where $A_i \cap A_j = \varnothing$ for all $i \neq j \in \mathbb{Z}_+$ and $A_i \in \mathcal{A}$ are subsets of a set $\mathcal{A}$, $\varnothing \in \mathcal{A}$ is an empty set-event. In the case when $\cup_{i\in\mathbb{Z}_+} A_i = \mathcal{A}$ then

$$p\left(\bigcup_{i\in\mathbb{Z}_+} A_i\right) = \sum_{i\in\mathbb{Z}_+} p(A_i) = 1\,. \tag{1.5}$$

This way, the first idea of linearity arises in relation to the problem of describing the motion of a particle in microworld, since we know that the quantum mechanical description of particle motion can be done in probabilistic terms. In particular, if a particle specified by a Lagrangian function $\mathcal{L}(\mathbf{x}, \dot{\mathbf{x}}; t)$, moves in microworld from a point $(\mathbf{x}, t)$ to a point $(\mathbf{x}', t') \in \mathbb{R}^3 \times \mathbb{R}_t$, this motion can be described only by measuring the probability $p(\mathbf{x}, t; \mathbf{x}', t') \geq 0$ of the event equivalent to propagating this particle along all possible trajectories $\mathbf{x}(s)$, $s \in [t, t']$, joining these two points:

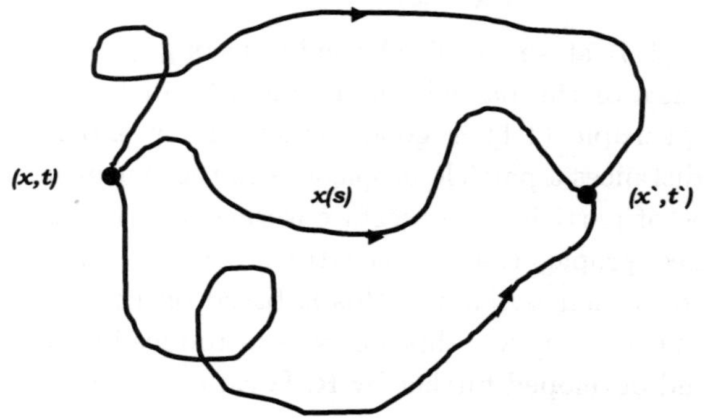

Fig. 1.1.

Thus, if we fix a point $(\mathbf{x}, t) \in \mathbb{R}^3 \times \mathbb{R}_t$, then evidently,

$$\sum_{\{\mathbf{x}'\}} p(\mathbf{x}, t; \mathbf{x}', t') = 1 . \tag{1.6}$$

The ingenious, prominent idea to define the probability functional $p(\mathbf{x}, t; \mathbf{x}', t') \geq 0$ satisfying (1.6) was an *amplitude* functional depending on each trajectory and belonging to some *linear space of functions.* Namely, let us consider some subsidiary linear functions space $\mathcal{H}$, containing the following elements or vectors built over every trajectory joining points $(\mathbf{x}, t)$ and $(\mathbf{x}', t')$:

$$A\{\mathbf{x}(\tau)|\mathbf{x}, t; \mathbf{x}', t'\} = \exp\left\{\frac{i}{\hbar}\int_{(\mathbf{x}',t')}^{(\mathbf{x},t)} \mathcal{L}(\mathbf{x}, \dot{\mathbf{x}}; \tau)\, d\tau\right\} \tag{1.7}$$

where $\hbar(\hbar = h/2\pi)$ is the Planck's constant, since its dimension is the same as that of the action functional (1.2). Thus, all vectors (1.7) belong to the linear space $\mathcal{H}$:

$$A \in \mathcal{H}\,, \tag{1.8}$$

meaning that if A_1 and $A_2 \in \mathcal{H}$, then $A_1 + A_2 \in \mathcal{H}$ too. Let us take now for instance, two different trajectories $\mathbf{x}_1(\tau)$ and $\mathbf{x}_2(\tau)$ joining points $(\mathbf{x}, t)$ and $(\mathbf{x}', t')$. Then, one can construct the amplitude

$$A = A_1 + A_2 = \exp\left\{\frac{i}{\hbar}\int_{(\mathbf{x}',t')}^{(\mathbf{x},t)} \mathcal{L}(\mathbf{x}_1, \dot{\mathbf{x}}_1; \tau)\, d\tau\right\} + \exp\left\{\frac{i}{\hbar}\int_{(\mathbf{x}',t')}^{(\mathbf{x},t)} \mathcal{L}(\mathbf{x}_2, \dot{\mathbf{x}}_2; \tau)\, d\tau\right\} \in \mathcal{H}\,. \tag{1.9}$$

By means of the procedure like (1.9) one can sum up all amplitudes from $(\mathbf{x}, t)$ to $(\mathbf{x}', t')$ corresponding to all trajectories joining these points:

$$\sum_{\substack{\{all\ trajectories\\ joining\ (\mathbf{x},t)\ and\ (\mathbf{x}',t')\}}} \exp\left\{\frac{i}{\hbar}\int_{(\mathbf{x}',t')}^{(\mathbf{x},t)} \mathcal{L}(\mathbf{x}, \dot{\mathbf{x}}; \tau)\, d\tau\right\} = \psi(\mathbf{x}, t; \mathbf{x}', t')\,. \tag{1.10}$$

We defined above a new element $\psi(\mathbf{x}, t; \mathbf{x}', t')$ pertaining evidently to the linear space $\mathcal{H}$, and which accumulates data of all possible paths reaching point $(\mathbf{x}, t)$ from point $(\mathbf{x}', t')$. Owing to probabilities reasoning mentioned above, the following crucial identification was made in early twenties of the last century:

$$p(\mathbf{x}, t; \mathbf{x}', t') = |\psi(\mathbf{x}, t; \mathbf{x}', t')|^2 d^3\mathbf{x}' . \tag{1.11}$$

The identification (1.11) clearly gives rise to the main property $p(\mathbf{x}, t; \mathbf{x}', t') \geq 0$ as $|\psi(\mathbf{x}, t; \mathbf{x}', t')|^2 \geq 0$ always. Further, based on the equality (1.6) the function (1.10) must satisfy the following constraint:

$$\sum_{\{\mathbf{x}'\}} |\psi(\mathbf{x}, t; \mathbf{x}', t')|^2 d^3\mathbf{x}' = 1 .$$

Replacing above the summation over $\mathbf{x}' \in \mathbb{R}^3$ by means of the usual integration, we can write it down as

$$\int_{\mathbb{R}^3} |\psi(\mathbf{x}, t; \mathbf{x}', t')|^2 d^3\mathbf{x}' = 1 \tag{1.12}$$

for all $\mathbf{x} \in \mathbb{R}^3$ and $t, t' \in \mathbb{R}_t$.

The function (1.10) is known also as a *wave function* completely characterizing the motion of a quantum particle in microworld. Returning to the definition (1.10), it is commonly used to write down this expression as the integral over paths:

$$\psi(\mathbf{x}, t; \mathbf{x}', t') = \int_{(\mathbf{x}', t')}^{(\mathbf{x}, t)} \mathcal{D}\{\mathbf{x}(\tau)\} \exp\left\{ \frac{i}{\hbar} \int_{(\mathbf{x}', t')}^{(\mathbf{x}, t)} \mathcal{L}(\mathbf{x}, \dot{\mathbf{x}}; \tau)\, d\tau \right\} , \tag{1.13}$$

this form was first introduced by R. Feynman when reinterpreting a result of P.A.M. Dirac about the canonical transformations in quantum mechanics.

As it can be shown easily enough, the function (1.13) enjoys the following dual to (1.12) relationship:

$$\int_{\mathbb{R}^3} |\psi(\mathbf{x}, t; \mathbf{x}', t')|^2 d^3\mathbf{x} = 1 \tag{1.14}$$

for all $\mathbf{x}' \in \mathbb{R}^3$ and $t, t' \in \mathbb{R}_t$; this is due to the symmetry between points (x, t) and (x', t'). Besides this, one can derive also an evolution equation governing the function (1.13) in time:

$$\left(-\frac{\hbar}{i}\right) \frac{\partial \psi}{\partial t} = \hat{H}\psi , \tag{1.15}$$

where, by definition an operator $\hat{H} : \mathcal{H} \to \mathcal{H}$ is defined as a linear generator of the infinitesimal shifting along the time variable $t \in \mathbb{R}_t$:

$$\hat{H}\psi = -\frac{\hbar}{i}\frac{d}{d\varepsilon}\psi(\mathbf{x}, t+\varepsilon; \mathbf{x}', t')|_{\varepsilon=0}\,. \tag{1.16}$$

Since the property (1.14) holds for all $(\mathbf{x}', t') \in \mathbb{R}^3 \times \mathbb{R}_t$, we need to take the linear functional space $\mathcal{H}$ to be a Hilbert space endowed with the usual scalar product of arbitrary vectors $\varphi, \psi \in \mathcal{H}$:

$$(\varphi, \psi) = \int\limits_{\mathbb{R}^3} \varphi^*(\mathbf{x})\psi(\mathbf{x})d^3\mathbf{x}\,. \tag{1.17}$$

Thereby, the operator $\hat{H}$ defined above being self-adjoint always, acts on vectors of the Hilbert space $\mathcal{H}$. This operator is called a Hamiltonian operator which is given in case when a Lagrangian

$$\mathcal{L} = \frac{1}{2}m|\dot{\mathbf{x}}|^2 - U(\mathbf{x}; t)\,, \tag{1.18}$$

as the differential operator:

$$\hat{H} = -\frac{\hbar^2}{2m}\Delta_{\mathbf{x}} + U(\mathbf{x}; t)\,, \tag{1.19}$$

where $\Delta_{\mathbf{x}} = \sum_{j=1}^{3} \frac{\partial^2}{\partial \mathbf{x}_j^2}$ is the standard Laplacian in three dimensions. Therefore, the equation (1.15) takes the form:

$$-\frac{\hbar}{i}\frac{\partial\psi}{\partial t} = -\frac{\hbar^2}{2m}\sum_{j=1}^{3}\frac{\partial^2\psi}{\partial \mathbf{x}_j^2} + U(\mathbf{x}; t)\psi\,, \tag{1.20}$$

known in quantum mechanics as the *Schrödinger equation.*

Similarly, one can find the operator $\hat{H}$, for instance, for a Lagrangian $\mathcal{L}$ describing the interaction of a charged particle with a magnetic field $\mathbf{B}$ expressed through a vector potential $\mathbf{A}$, where $\mathbf{B} = \operatorname{rot}\mathbf{A}$. In this case one has the Lagrangian

$$\mathcal{L} = \frac{1}{2}m|\dot{\mathbf{x}}|^2 - e < \mathbf{A}, \dot{\mathbf{x}} > - U(\mathbf{x}; t)\,, \tag{1.21}$$

which produces the following Hamiltonian operator:

$$\hat{H} = \frac{1}{2m}\left|\frac{\hbar}{i}\frac{\partial}{\partial \mathbf{x}} + e\mathbf{A}\right|^2 + U(\mathbf{x}; t)\,. \tag{1.22}$$

Now we are almost in a position to go over to introducing a general notion of quantum physics called a state. We saw above that evolution of a particle in microworld can be specified by means of an amplitude wave vector $\psi(\mathbf{x}, t; \mathbf{x}', t') \in \mathcal{H}$ which is completely characterized by the Schrödinger equation (1.15) predicting the future state of the particle. Based on the reasoning like this any normalized vector $\psi \in \mathcal{H}$ satisfying the equation (1.15) can be called by a dynamical state of our quantum object in microworld $\mathbb{R}^3 \times \mathbb{R}_t$. Since the equation (1.15) is linear in a Hilbert space $\mathcal{H}$, one can construct so called stationary states corresponding to the spectrum of the operator $\hat{H} : \mathcal{H} \to \mathcal{H}$. If $\psi_E \in \mathcal{H}$ satisfies the equation (1.15) with an eigenvalue $E \in \mathbb{R}$, that is

$$\left(-\frac{\hbar}{i}\right) \frac{\partial \psi_E}{\partial t} = \hat{H}\psi_E = E\psi_E\,, \tag{1.23}$$

whence one finds that

$$\psi_E = \bar{\psi}_E(\mathbf{x}) \exp\left(-\frac{i}{\hbar}E\right), \tag{1.24}$$

where $\bar{\psi}_E(\mathbf{x})$ does not depend on $t \in \mathbb{R}$. The states (1.24) are called stationary states of our quantum system being of great importance for many applications. We now proceed to one of these applications.

Chapter 2

Quantum States and Statistics

Let us consider a one particle quantum dynamical system specified by a Hamiltonian operator $\hat{H}$ acting on Hilbert space $\mathcal{H}$ endowed with a scalar product $\langle\beta|\alpha\rangle$. Its quantum stationary state which we will denote, due to P. Dirac, as $|\alpha\rangle \in \mathcal{H}$, satisfies the Schrödinger equation:

$$\hat{H}|\alpha\rangle = \left(-\frac{\hbar}{i}\right)\frac{\partial}{\partial t}|\alpha > \Longrightarrow E_\alpha|\,\alpha\rangle\,, \tag{2.1}$$

where $E_\alpha \in \mathbb{R}$ is the corresponding eigenvalue of the operator $\hat{H} : \mathcal{H} \to \mathcal{H}$. It is now interesting to consider a system of the same one-particle states, so called an ensemble, not interacting with each other. Due to Pauli and Fermi, this ensemble of $N \in \mathbb{Z}_+$ identical particles should evidently be specified by a vector of state belonging now to the tensor product of $N \in \mathbb{Z}_+$ Hilbert spaces $\mathcal{H}$. This can be written down as

$$|\alpha_N\rangle = \bigotimes_{j=1}^{N}|\alpha_j\rangle \in \bigotimes_{j=1}^{N}\mathcal{H}\,. \tag{2.2}$$

But the state (2.2) still distinguishes particles in the ensemble in spite of the natural assumption that identical particles are indistinguishable. Due to experimental facts it was stated that all N-particle states like (2.2) can either be symmetrical or anti-symmetrical subject to any particle permutation inside an ensemble. This fact motivates one to derive from (2.2) both symmetrical and anti-symmetrical states via the standard (anti)-symmetrization:

$$\begin{aligned}|(\alpha)_N\rangle &\Longrightarrow |\alpha_1, \alpha_2, \alpha_3, \cdots, \alpha_N\ \rangle \\ &= \frac{1}{(N!)^{1/2}}\sum_{\sigma\in S_N}\varepsilon(\sigma)|\alpha_{\sigma(1)} > \otimes|\alpha_{\sigma(2)}\rangle \otimes \cdots \otimes |\alpha_{\sigma(N)}\rangle\,,\end{aligned} \tag{2.3}$$

where $\varepsilon(\sigma) = \pm 1$, $\otimes$-sign means the usual tensor product of vectors and the coefficient $(N!)^{-1/2}$ is added concerning the normalization of the state (2.3) by unity if all states $|\alpha_j> \in \mathcal{H}$, $j = \overline{1, N}$, are normalized by unity too. One can see that the vector state (2.3) is either symmetric or antisymmetric subject to a particle permutation inside an ensemble. (Here S_N denotes all possible permutations $\sigma \in S_N$ of N integers $\{1, 2, \cdots, N\}$).

If we take any other state vector $|(\beta)_N> = |\beta_1, \beta_2, \cdots, \beta_N> \in \bigotimes_{sym} \mathcal{H}^N$, then their scalar product can naturally be defined as

$$\langle \beta_N, \beta_{N-1}, \cdots, \beta_1 | \alpha_1, \alpha_2, \cdots, \alpha_N \rangle = |\langle \beta_j | \alpha_k \rangle|^N_{\sigma, j, k=1} ,$$
$$\langle \beta_j | \alpha_k \rangle = \int_{\mathbb{R}^3} \beta_j^*(\mathbf{x}) \alpha_k(\mathbf{x}) d^3\mathbf{x} , \tag{2.4}$$

where the right hand side of (2.4) coincides either with $det(\cdots)$ if $\varepsilon(\sigma) = \pm 1$, or with $per(\cdots)$, if $\varepsilon(\sigma) \equiv 1$. These two cases are called Fermi-antisymmetric one, and Bose-symmetric one. Correspondingly, particles in these ensembles are called fermions and bosons. It is natural to call the physical state $|0> \in \mathcal{H}$ describing the empty ensemble *a vacuum*.

From the above construction one can easily write down a quantum state consisting of any $k-$particle ensemble:

$$|(\beta)\rangle = \beta_0 |0\rangle \bigoplus |\beta_1^{(1)}\rangle \bigoplus |\beta_1^{(2)}, \beta_2^{(2)}\rangle \bigoplus \cdots$$
$$\bigoplus |\beta_1^{(N)}, \beta_2^{(N)}, \cdots, \beta_{N-1}^{(N)}, \beta_N^{(N)}\rangle , \tag{2.5}$$

where $|\beta_s^{(j)}\rangle$, $s = \overline{1, j}$, are the corresponding one particle states of $j-$th ensemble, $j = \overline{1, N}$. The expression (2.5) evidently means that the ambient Hilbert space $\oplus$ containing all vectors like (2.5) can be represented as

$$\Phi = \bigoplus_{j \in Z_+} \Phi_j , \tag{2.6}$$

< where $\Phi_j = \bigotimes_{\{sym/antisym\}} \mathcal{H}^j$, $j = \overline{1, N}$, are, as mentioned above, (anti)-symmetrized Hilbert spaces

$\mathcal{H}$ of one particle states. This combined space Φ (2.6) is called *a Fock space.*

Chapter 3

Creation and Annihilation Operators

Let us consider an arbitrary one-particle state $|\beta\rangle \in \Phi_1 \simeq \mathcal{H}$ and define naturally a linear operation $a^+(\beta)$ of adding this state $|\beta\rangle$ to any other state of the Fock space Φ as follows:

$$a^+(\beta)|\alpha_1, \alpha_2, \cdots, \alpha_N\,\rangle = |\beta, \alpha_1, \alpha_2, \cdots, \alpha_N\rangle\,, \tag{3.1}$$

where taking

$$|\alpha_1, \alpha_2, \cdots, \alpha_N\,\rangle \in \Phi_N \tag{3.2}$$

gives the vector

$$|\beta, \alpha_1, \alpha_2, \cdots, \alpha_N\,\rangle \in \Phi_{N+1}. \tag{3.3}$$

When $N = 0$, (3.1) leads us

$$a(\beta)|0\,\rangle = |\beta\rangle \in \Phi_1\,. \tag{3.4}$$

It is so reasonable to call the operator $a(\beta) : \Phi \to \Phi$ the annihilation (or destruction) operator, since

$$a(\beta)|\alpha_1, \alpha_2, \cdots, \alpha_N\rangle = \sum_{j=1}^{N} \varepsilon^{j-1}\langle\beta|\alpha_j\rangle|\alpha_1, \alpha_2, , \hat{\alpha}_j, , \alpha_{N-1}, \alpha_N\rangle \in \Phi_{N-1}\,, \tag{3.5}$$

where "$\hat{\alpha}_j$", $j = \overline{1, N}$, means that these states should be omitted. Making now the use of definition (3.4) and the property (3.5) (being derived!), one can see easily that the following commutation relations hold:

$$\begin{gathered}[a^+(\beta_j), a^+(\beta_k)]_{-\varepsilon} = 0 = [a(\beta_j), a(\beta_k)]_{-\varepsilon}\,, \\ [a(\beta_j), a^+(\beta_k)]_{-\varepsilon} = \langle\beta_j|\beta_k\rangle\end{gathered} \tag{3.6}$$

for all $\beta_j \in \Phi_1 \simeq \mathcal{H}$, $j,k = \overline{1,N}$, where $[\cdots,\cdots]_{-\varepsilon}$ means commutator for $\varepsilon = 1$, and anticommutator for $\varepsilon = -1$.

Consider now a linear operator $A : \Phi \longrightarrow \Phi$ acting only on one-particle states of Φ, i.e.

$$A|\alpha_1,\alpha_2,\cdots,\alpha_N\rangle = \sum_{j=1}^{N} |\alpha_1,\alpha_2,\cdots \left(A^{(1)}\alpha_j\right),\cdots,\alpha_N\,\rangle\,, \tag{3.7}$$

where we denoted $A|_{\Phi_1}$ by $A^{(1)}$: that is *the projection* of A onto the subspace Φ_1, which is assumed to be *invariant*! It is possible to represent this operator A by means of the basic operators $a^+(\beta)$ and $a(\beta)$ as follows: consider any operator expression acting on the one-particle state $\Phi_1 \simeq \mathcal{H}$ in the following well known tensor form:

$$A^{(1)}_{\beta\gamma} = |\beta\rangle\langle\gamma| \colon \Phi_1 \longrightarrow \Phi_1\,, \tag{3.8}$$

where by definition, for any one-particle state $|\alpha\rangle \in \Phi$

$$(|\beta\rangle\langle\gamma|)|\alpha\rangle = \langle\gamma|\alpha\rangle|\beta\rangle \in \Phi_1\,. \tag{3.9}$$

Using (3.8) we can now calculate the expression (3.6) that gives rise to the following result:

$$\begin{aligned}
&A_{\beta\gamma}|\alpha_1,\alpha_2,\cdots,\alpha_N\rangle \\
&\quad= \sum_{j=1}^{N} |\alpha_1,\alpha_2,\cdots,A^{(1)}_{\beta\gamma}\alpha_j,\cdots,\alpha_N\rangle \\
&\quad= \sum_{j=1}^{N} |\alpha_1,\alpha_2,\cdots,|\beta\rangle\langle\gamma|\alpha_j\rangle,\cdots,\alpha_N\rangle \\
&\quad= \sum_{j=1}^{N} (-1)^j\langle\gamma|\alpha_j\rangle \mid \beta,\alpha_1,\alpha_2,\cdots,\alpha_j,\alpha_{j+1},\cdots,\alpha_N\rangle \\
&\quad= \sum_{j=1}^{N} (-1)^j\langle\gamma|\alpha_j\rangle a^+(\beta)|\alpha_1,\alpha_2,\cdots,\hat{\alpha}_j,\cdots,\alpha_N\rangle \\
&\quad= a^+(\beta)\sum_{j=1}^{N} (-1)^j\langle\gamma|\alpha_j\rangle|\alpha_1,\alpha_2,\cdots,\hat{\alpha}_j,\cdots,\alpha_N\rangle \\
&\quad\equiv a^+(\beta)a(\gamma)|\alpha_1,\alpha_2,\cdots,\alpha_N\rangle
\end{aligned} \tag{3.10}$$

Thus, we can identify two operators: $A_{\beta\gamma}$ and $a^+(\beta)a(\gamma)$ because their *actions* are the same; that is

$$A_{\beta\gamma} = a^+(\beta)a(\gamma)\,. \tag{3.11}$$

It is useful to consider any complete orthogonal system of one-particle states $|\alpha_j\rangle \in \Phi_1$, $j \in \mathbb{Z}_+$, satisfying the conditions

$$\sum_{j\in\mathbb{Z}_+} |\alpha_j\rangle\langle\alpha_j| = 1, \langle\alpha_j|\alpha_k\ \rangle = \delta_{jk} \tag{3.12}$$

for all $j, k \in \mathbb{Z}_+$. This means for instance, that any operator $A^{(1)}\colon \Phi_1 \longrightarrow \Phi_1$ can be represented as

$$A^{(1)} = \sum_{j,k\in\mathbb{Z}_+} A^{(1)}_{jk}|\alpha_j\ \rangle\langle\ \alpha_k|\,. \tag{3.13}$$

When extended to the whole Fock space Φ, the operator (3.13) takes the form:

$$A \Longrightarrow \sum_{j,k\in\mathbb{Z}_+} A^{(1)}_{jk} a^+(\alpha_j)a(\alpha_k)\,, \tag{3.14}$$

where, due to (3.12), the following *canonical* relationships:

$$[a_j, a_k^+]_{-\varepsilon} = \delta_{jk}\,. \tag{3.15}$$

hold for operators $a_j^+ := a^+(\alpha_j)$ and $a_k := a(\alpha_k)$, $j, k \in \mathbb{Z}_+$. For the coefficients $A^{(1)}_{jk}$, $j, k \in \mathbb{Z}_+$ in (3.13), it is easy to find that

$$A^{(1)}_{jk} = \langle\alpha_j|A^{(1)}|\alpha_k\rangle\,, \tag{3.16}$$

are the so called "matrix" elements of the operator $A^{(1)}$ in Φ with respect to an orthogonal complete basis of one particle states $|\alpha_j\rangle \in \Phi_1$, $j \in \mathbb{Z}_+$.

As an example, let us consider the particle number operator $\hat{N}$,representing the of number of particles in a state $|\alpha_1, \cdots, \alpha_N\rangle \in \Phi$.Then,

$$\hat{N}|_{\Phi_1} = \mathbb{I} \tag{3.17}$$

is the identity operator. Subject to a complete orthonormal basis of one-particle states, we can easily find that

$$\hat{N}_{jk} = \langle\ \alpha_j|\mathbb{I}|\alpha_k\ \rangle = \langle\ \alpha_j|\alpha_k\ \rangle = \delta_{jk}\,. \tag{3.18}$$

for all $j, k \in \mathbb{Z}_+$. Thus, owing to the expression (3.14), one arrives at

$$\hat{N} \Longrightarrow \sum_{j,k\in\mathbb{Z}_+} \delta_{jk}\, a_j^+ a_k = \sum_{j\in\mathbb{Z}_+} a_j^+ a_j \,. \tag{3.19}$$

This procedure can be generalized to the case of operators $\hat{A} : \Phi \longrightarrow \Phi$ acting *effectively only* on two-particle states $|\alpha_1, \alpha_2\ \rangle \in \Phi_2$, giving rise to operators like

$$A \Longrightarrow \sum_{j,k,s,l\in\mathbb{Z}_+} A^{(2)}_{jk,sl} a_j^+ a_k^+ a_s a_l \,, \tag{3.20}$$

where as before, $A^{(2)}_{jk,sl}$, j, k and $s, l \in \mathbb{Z}_+$, are the corresponding matrix elements of the operator $A^{(2)} : \Phi_2 \longrightarrow \Phi_2$:

$$A^{(2)}_{jk,sl} = \langle\ \alpha_j | \langle\ \alpha_k | A^{(2)} | \alpha_s\ \rangle | \alpha_l\ \rangle \tag{3.21}$$

subject to a complete orthogonal system of one-particle states $|\alpha_j\ \rangle \in \Phi_1$, $j \in \mathbb{Z}_+$.

The representations (3.16) and (3.20) constructed above are called *second quantized* in quantum field theory, and the procedure of getting expressions like these is called *second quantization.*

Chapter 4

Some Second Quantized Quantum Models of Interacting Particles and Fields (Bosons and Fermions)

Model 1

Let us now consider the simplest case when a Hamiltonian $\hat{H}$ in a Fock space Φ is stemming from a two-particle additive interaction like

$$\hat{H}_{int}^{(2)}(\mathbf{x},\mathbf{y}) = U^{(2)}(\mathbf{x}-\mathbf{y}) = U^{(2)}(\mathbf{y}-\mathbf{x}) \tag{4.1}$$

where $\mathbf{x},\mathbf{y} \in \mathbb{R}^3$ are positions of particles in space $\mathbb{R}^3$. This means, that any $N-$particle system of such particles is described by a Hamiltonian $\hat{H}_N : \Phi_N \longrightarrow \Phi_N$, where due to (1.20)

$$\hat{H}_N = \sum_{j=1}^{N} -\frac{\hbar^2}{2m}\nabla^2_{\mathbf{x}_j} + \frac{1}{2}\sum_{\substack{i,j=1\\ i\neq j}}^{N} U^{(2)}(\mathbf{x}_i,\mathbf{x}_j)\,. \tag{4.2}$$

Based on rules of the preceeding chapter 3, we can get that

$$\hat{H} = \int_{\mathbb{R}^3} \mathbf{d}^3\mathbf{x}\frac{\hbar^2}{2m}\langle\ \nabla_{\mathbf{x}}\hat{\psi}^+(\mathbf{x}), \nabla_{\mathbf{x}}\hat{\psi}(\mathbf{x})\ \rangle$$

$$+\frac{1}{2}\int_{\mathbb{R}^3}\int_{\mathbb{R}^3} d^3\mathbf{x}d^3\mathbf{y}\ \hat{\psi}^+(\mathbf{x})\hat{\psi}^+(\mathbf{y})U^{(2)}(\mathbf{x},\mathbf{y})\hat{\psi}(\mathbf{y})\hat{\psi}(\mathbf{x})\,, \tag{4.3}$$

where we have used the so called "coordinate representation" for operators $a^+(\alpha)$ and $a(\alpha)$:

$$a^+(\alpha) = a^+(\alpha(\mathbf{x})) = \hat{\psi}^+(\mathbf{x}),\ a(\alpha) = a(\alpha(\mathbf{x})) = \hat{\psi}(\mathbf{x})\,, \tag{4.4}$$

and where $|\alpha\rangle \Rightarrow |\alpha(\mathbf{x})\ \rangle \in \Phi_1$ is a state with a fixed $\mathbf{x}$-coordinate. Due to the representation (3.24) one has for instance, that

$$\hat{N} = \int_{\mathbb{R}^3} d^3\mathbf{x}\hat{\psi}^+(\mathbf{x})\hat{\psi}(\mathbf{x})\,, \tag{4.5}$$

that is the operator $\hat{N} = \hat{\psi}^+(\mathbf{x})\hat{\psi}(\mathbf{x})$ has an interpretation as the density operator for particles in the Fock space Φ.

If we make use of the "momentum" representation for states $|\alpha\rangle \Rightarrow |\alpha_{\mathbf{k}}\rangle \in \Phi_1$, where $\mathbf{k} \in \mathbb{R}^3$, $|\mathbf{k}| \leq \left(\frac{2\pi}{V}\right)^3$, is the momentum of a particle in the state $|\alpha_{\mathbf{k}}\rangle \in \Phi_1$, then similarly one arrives at the expression

$$\hat{H} = \sum_{\mathbf{k}} \frac{\hbar^2|\mathbf{k}|^2}{2m} a_{\mathbf{k}}^+ a_{\mathbf{k}} + \frac{1}{2} \sum_{\mathbf{k},\mathbf{s},\mathbf{l},\mathbf{r}} a_{\mathbf{k}}^+ a_{\mathbf{s}}^+ U_{\mathbf{ks},\mathbf{lr}}^{(2)} a_{\mathbf{l}} a_{\mathbf{r}} \,, \tag{4.6}$$

where, by definition, $|\alpha_{\mathbf{k}}\rangle$-states are normalized by unity, $a_{\mathbf{k}} = a(\alpha_{\mathbf{k}})$, $a_{\mathbf{k}}^+ = a^+(\alpha_{\mathbf{k}})$,

$$\langle \alpha_{\mathbf{k}} | \alpha_{\mathbf{s}} \rangle = \delta_{\mathbf{ks}} \,, \tag{4.7}$$

and as before, states $|\alpha_{\mathbf{k}}\rangle \in \Phi_1$, $\mathbf{k} \in \mathbb{R}^3, |\mathbf{k}| \leq \left(\frac{2\pi}{V}\right)^3$, enjoy the one-particle Schrödinger equation:

$$\hat{H}^{(1)}|\alpha_{\mathbf{k}}\rangle = \frac{|\mathbf{k}|^2}{2m}|\alpha_{\mathbf{k}}\rangle, \; |\alpha_{\mathbf{k}}\rangle = \frac{1}{\sqrt{V}} e^{i\langle k,\mathbf{x}\rangle} \,, \tag{4.8}$$

with energy eigenvalues $\varepsilon_{\mathbf{k}} = \frac{|\mathbf{k}|^2}{2m}$. The momentum parameter can only be taken as a counting states parameter here due to the fact that the momentum operator $\hat{P} : \Phi \longrightarrow \Phi$ is a constant of motion for (4.6), that is

$$[\hat{P}, \hat{H}] = 0 \,, \tag{4.9}$$

where

$$\hat{P} = \sum_{\mathbf{k}} \hbar \mathbf{k} \tag{4.10}$$

This stems from the following form of the momentum operator $\hat{P}^{(1)}$ in $\mathbf{x}$-representation:

$$\hat{P}^{(1)} = \frac{\hbar}{i} \nabla_{\mathbf{x}} : \Phi_1 \longrightarrow \Phi_1 \,, \tag{4.11}$$

resulting due to the rule (3.25) in the following expression for $\hat{P}$: $\Phi \longrightarrow \Phi$:

$$\hat{P} = \frac{\hbar}{2i} \int_{\mathbb{R}^3} d^3\mathbf{x} \left(\nabla_{\mathbf{x}} \hat{\psi}^+(\mathbf{x}) \hat{\psi}(\mathbf{x}) - \hat{\psi}^+(\mathbf{x}) \nabla_{\mathbf{x}} \hat{\psi}(\mathbf{x}) \right) \tag{4.12}$$

Here we have used the symmetrization trick for getting a self-adjoint expression.

The condition (4.9) clearly means that, states $|\alpha_{\mathbf{k}}\rangle \in \Phi_1$ are eigen-states of operators $\hat{P}$ and $\hat{H}$ simultaneously; this is obvious since

$$\left[-\frac{\hbar^2}{2m}\nabla^2_{\mathbf{x}},\ \frac{\hbar}{i}\nabla_{\mathbf{x}}\right] \equiv 0\,. \tag{4.13}$$

Model 2

In the case when there exists another physically important operator commuting with the Hamiltonian (4.3), it is possible to add the corresponding indices to states $|\alpha_{\mathbf{k}}\rangle \in \Phi_1$ marking the dependence on its eigenvalues. For example, if we take the spin operator $\sigma : \Phi \longrightarrow \Phi$, consisting of Pauli matrices, we obviously get

$$[\sigma, \hat{H}] = 0\,. \tag{4.14}$$

Since the spin-operator $\sigma\colon \Phi_1 \longrightarrow \Phi_1$ possesses only *two* eigenvalues, $\hbar/2$ and $-\hbar/2$ in the case of fermions (like electrons), this means that we should specify our eigen-state $|\alpha_{\mathbf{k}}\rangle \in \Phi_1$ with an additional parameter, for instance $\gamma \in \{-1, +1\}$, getting the state $|\alpha_{\mathbf{k},\gamma}\rangle \in \Phi_1$. Thereby, we can now say, that a state $|\alpha_{\mathbf{k},+1}\rangle \in \Phi_1$ describes a fermion particle with momentum $\mathbf{k} \in \mathbb{R}^3$ and spin $+\hbar/2$ directed upward, and similarly, that a state $|\alpha_{\mathbf{k},-1}\rangle \in \Phi_1$ describes a fermion particle with momentum $\mathbf{k} \in \mathbb{R}^3$ and spin $-\hbar/2$ directed downward.

As in above, if spin possessing particles interact with each other, we get the following Hamiltonian $\hat{H}^{(2)}_{int}$ (similarly to the expression (4.1)):

$$\hat{H}^{(2)}_{int}(\mathbf{x},\mathbf{y}) = U^{(2)}_0(\mathbf{x}-\mathbf{y})(\mathbb{I}(\mathbf{x})\mathbb{I}(\mathbf{y})) + U^{(2)}_1(\mathbf{x}-\mathbf{y})\langle\, \sigma(\mathbf{x})\otimes, \sigma(\mathbf{y})\,\rangle\,, \tag{4.15}$$

here $\sigma(\mathbf{x}) : \Phi_1 \longrightarrow \Phi_1$ is the spin operator of a particle at point $\mathbf{x} \in \mathbb{R}^3$, and $\sigma(\mathbf{y}) : \Phi_1 \longrightarrow \Phi_1$ is that at point $\mathbf{y} \in \mathbb{R}^3$, $\langle\, \cdot, \cdot\,\rangle$ is the usual scalar product in $\mathbb{R}^3$ and

$$\sigma(\mathbf{x}) = \sigma_1\vec{\imath} + \sigma_2\vec{\jmath} + \sigma_3\vec{k} \tag{4.16}$$

with Pauli matrices in the form

$$\sigma_1 = \begin{bmatrix} 0 & 1 \\ 1 & 0 \end{bmatrix},\quad \sigma_2 = \begin{bmatrix} 0 & i \\ -i & 0 \end{bmatrix},\quad \sigma_3 = \begin{bmatrix} 1 & 0 \\ 0 & -1 \end{bmatrix}. \tag{4.17}$$

Model 3

As another example let us consider an ensemble of spin-particles in a volume $V \subset \mathbb{R}^3$ interacting with an external magnetic field $\mathbf{B}$ As is well known from *Electricity* course, one-particle interaction energy $\hat{H}_{int}$ with a magnetic field is equal to the expression like

$$\hat{H}_{int}(\mathbf{x}) = -\mu_0 \langle\, \sigma(\mathbf{x}),\ \mathbf{B}(\mathbf{x})\, \rangle\,, \tag{4.18}$$

where spin $\sigma(\mathbf{x})$ here is interpreted in the same way as a magnetic dipole $\mu(\mathbf{x})$ interacting with a magnetic field. Since the expression (4.18) acts only on one-particles states of Φ_1, following the rules of Chapter 3 one derives from (4.18) the second quantized form

$$\hat{H}_{int} \Longrightarrow -\mu_0 \int\limits_{\mathbb{R}^3} d^3\mathbf{x} \sum_{\gamma,\mu} \hat{\psi}^+_\gamma(\mathbf{x}) \langle\, \sigma_{\gamma\mu},\ \mathbf{B}(\mathbf{x})\, \rangle \hat{\psi}_\mu(\mathbf{x})\,, \tag{4.19}$$

or the momentum-representation mentioned before,

$$\hat{H}_{int} = -\mu_0 \sum_{\mathbf{k},\mathbf{s},\gamma,\mu} \langle\, \sigma_{\gamma\mu},\ \mathbf{B}_{\mathbf{ks}}\, \rangle a^+_{\mathbf{k},\gamma} a_{\mathbf{s},\mu}\,, \tag{4.20}$$

where $\mathbf{B}_{\mathbf{ks}}$ are the corresponding matrix elements and $\mu_0 \in \mathbb{R}_+$ is the Bohr magneton. Similarly, if one considers a system of charged particles distributed continuously, forming a current density $\mathbf{J}(\mathbf{x})$ at a point $\mathbf{x} \in \mathbb{R}^3$, and interacting with an external electromagnetic field $\mathbf{A}(\mathbf{x})$, the interaction $\hat{H}_{int}$ will take the form

$$\hat{H}^{(1)}_{int} = -\langle\, \mathbf{J}(\mathbf{x}), \mathbf{A}(\mathbf{x})\, \rangle\,, \tag{4.21}$$

or in the secondly quantized form

$$\hat{H}_{int} = -\int\limits_{\mathbb{R}^3} d^3\mathbf{x}\ \hat{\psi}^+(\mathbf{x}) \langle\, \mathbf{J}(\mathbf{x}), \mathbf{A}(\mathbf{x})\, \rangle \hat{\psi}(\mathbf{x})\,. \tag{4.22}$$

Model 4

Here, we remain concerned with the structure of the interaction Hamiltonian of electrons with crystal excitations called phonons. As is well known, the interaction mechanism is due to the *polarization* of the medium produced by *own lattice vibrations*. As a result of the polarization, the interaction energy of the free electrons inside the crystal is given by the Coulomb like potential form due to the crystal polarization:

$$d\hat{H}^{(2)}_{int}(\mathbf{x}, \mathbf{x}') = \frac{-ZeN\hat{\rho}(\mathbf{x}) d^3\mathbf{x}\ \mathrm{div}\mathbf{q}(\mathbf{x}') d^3\mathbf{x}'}{V|\mathbf{x} - \mathbf{x}'|}\,, \tag{4.23}$$

where $\mathbf{q}(\mathbf{x}) \in \mathbb{R}^3$ is a charge displacement vector. It is convenient to expand the displacement operator $\mathbf{q}(\mathbf{x}) \in \mathbb{R}^3$ in a Fourier series like

$$\mathbf{q}(\mathbf{x}) = \frac{-i}{(MN/V)^{1/2}} \sum_{\mathbf{k}} \left(\frac{\hbar}{2\omega_{\mathbf{k}} V} \right)^{1/2} \frac{\mathbf{k}}{|\mathbf{k}|} [c_{\mathbf{k}} \exp(i\langle\, \mathbf{k}, \mathbf{x}\,\rangle)$$
$$- i c_{\mathbf{k}}^{+} \exp(-i\langle\, \mathbf{k}, \mathbf{x}\,\rangle)] \vartheta(\omega_D - \omega_{\mathbf{k}})\,, \tag{4.24}$$

where $\omega_D = u_0 (6\pi^2 N/V)^{1/3}$ is the Debye frequency, u_0 is the velocity of sound in the crystal, and M is the mass of an ion. Having now applied the second quantization rules of Chapter 3, one arrives at the expression

$$\hat{H}_{int} = \frac{Ze^2}{u_0} \left(\frac{N}{VM} \right)^{1/2} \sum_{\mathbf{k},\gamma} \sum_{\mathbf{q}} \frac{4\pi}{|\mathbf{q}|^2} \vartheta(\omega_D - \omega_{\mathbf{k}})$$
$$\times [a_{\mathbf{k}+\mathbf{q},\gamma}^{+} a_{\mathbf{k},\gamma} c_{\mathbf{q}} + a_{\mathbf{k},\gamma}^{+} a_{\mathbf{k}+\mathbf{q},\gamma} c_{\mathbf{q}}^{+}]\,, \tag{4.25}$$

where $a_{\mathbf{k},\gamma}^{+}$ and $a_{\mathbf{s},\mu}$, $\mathbf{k}, \mathbf{s} \in \mathbb{R}$, are electron creation and destruction operators with spin variables $\gamma, \mu \in \{\pm 1\}$. The expression (4.25) can be rewritten down in the following convenient form if the factor $4\pi/|\mathbf{q}|^2$ is replaced by a constant like $4\pi/q_{TF}^2$, where $q_{TF}^2 = 4k_F/\pi a_0$:

$$\hat{H}_{int} = g \int_{\mathbb{R}^3} d^3\mathbf{x}\ \hat{\psi}_{\alpha}^{+}(\mathbf{x}) \hat{\psi}_{\alpha}(\mathbf{x}) \hat{\varphi}(\mathbf{x})\,, \tag{4.26}$$

where $g \simeq \dfrac{4\pi Z N (N/VM)^{1/2}}{u_0 q_{TF}^2}$,

$$\hat{\varphi}(\mathbf{x}) = \sum_{\mathbf{k}} \left(\frac{\hbar\omega_{\mathbf{k}}}{2V} \right) [c_{\mathbf{k}} \exp(i\langle\, \mathbf{k}, \mathbf{x}\,\rangle) + c_{\mathbf{k}}^{+} \exp(i\langle\, \mathbf{k}, \mathbf{x}\,\rangle)] \vartheta(\omega_D - \omega_{\mathbf{k}})\,, \tag{4.27}$$

and for all $\mathbf{k}, \mathbf{k}' \in \mathbb{R}^3$ the Bose commutation relationships hold:

$$[c_{\mathbf{k}}, c_{\mathbf{k}}^{+}] = \delta_{\mathbf{k}\mathbf{k}'}\,. \tag{4.28}$$

Concerning the phonon energy of a crystal it is convenient to make use of the harmonic oscillation energy expression:

$$\hat{H}_{ph} = \frac{1}{2} \rho \int_{\mathbb{R}^3} d^3\mathbf{x} \left\langle \frac{d\mathbf{q}}{dt}, \frac{d\mathbf{q}}{dt} \right\rangle + \sum_{\mathbf{q}} \hbar\omega_{\mathbf{q}} (c_{\mathbf{q}}^{+} c_{\mathbf{q}} + 1/2)\,, \tag{4.29}$$

useful enough together with expressions (4.25) and (4.6) for applications. The resulting Hamiltonian

$$\hat{H}_{el-ph} = \sum_{\mathbf{q}} \hbar\omega_{\mathbf{k}} \left(c^{+}_{\mathbf{q}} c_{\mathbf{q}} + \frac{1}{2} \right) + \sum_{\mathbf{k},\gamma} \frac{\hbar^2 |\mathbf{k}|^2}{2m} a^{+}_{\mathbf{k},\gamma} a_{\mathbf{k},\gamma}$$
$$+ \frac{1}{2V} \sum_{\mathbf{k},\mathbf{s},\mathbf{q},\gamma,\mu} \frac{4\pi}{\mathbf{q}^2} a^{+}_{\mathbf{k}+\mathbf{q},\gamma} a^{+}_{\mathbf{s}-\mathbf{q},\mu} a_{\mathbf{s},\mu} a_{\mathbf{k},\gamma}$$
$$+ g \sum_{\mathbf{k}} \frac{\vartheta(\omega_D - \omega_{\mathbf{q}})}{V} [a^{+}_{\mathbf{k}+\mathbf{q},\gamma} a_{\mathbf{k},\gamma} c_{\mathbf{q}} + a^{+}_{\mathbf{k},\gamma} a_{\mathbf{k}+\mathbf{q},\gamma} c^{+}_{\mathbf{q}}] \tag{4.30}$$

is known as the Fröhlich Hamiltonian in Quantum Physics and is of fundamental importance particularly when describing the super-conductivity at low temperatures.

Chapter 5

The Rayleigh-Schrödinger Approach to the Ground State Spectrum Calculations

Let us consider a quantum system which is specified by a Hamiltonian operator $\hat{H} : \Phi \longrightarrow \Phi$

$$\begin{aligned}\hat{H} &= \sum_{\mathbf{k}} \varepsilon_{\mathbf{k}} a^+_{\mathbf{k}} a_{\mathbf{k}} + \sum_{\mathbf{k,s,q}} U^{(2)}_{\mathbf{k,s,q}} a^+_{\mathbf{k}} a^+_{\mathbf{s}} a_{\mathbf{s+q}} a_{\mathbf{s-q}} \\ &\quad + i \sum_{\mathbf{k,s,q}} U^{(1,+)}_{\mathbf{k,s,q}} (a^+_{\mathbf{k}} a^+_{\mathbf{s}} b_{\mathbf{q}} - a_{\mathbf{s}} a_{\mathbf{k}} b^+_{\mathbf{q}}) \\ &\quad + i \sum_{\mathbf{k,s,q}} U^{(1,-)}_{\mathbf{k,s,q}} (a^+_{\mathbf{k}} a_{\mathbf{s}} b_{\mathbf{q}} - a^+_{\mathbf{s}} a_{\mathbf{k}} b^+_{\mathbf{q}}) \\ &= \hat{H}_0 + \hat{H}_{int} \,, \end{aligned} \tag{5.1}$$

where $U^{(1,+)}, U^{(1,-)}$ and $U^{(2)}$ are the corresponding interaction potentials. As we know the particle number operator $\hat{N} = \sum_{\mathbf{k}} a^+_{\mathbf{k}} a_{\mathbf{k}}$ is in general not commuting with Hamiltonian (1.1), i.e.

$$[\hat{H}, \hat{N}] \neq 0 \tag{5.2}$$

that means (5.1) describes the particle nonconservative nature of the physical process. Namely, due to the Heisenberg equations for evolution of operators, one arrives at

$$\frac{d\hat{N}}{dt} = \frac{i}{\hbar}[\hat{H}, \hat{N}] \,. \tag{5.3}$$

In particular, the Hamiltonian (5.1) can describe a process of creating new particles in the system.

If a closed system is in the state of equilibrium, say in the state $|\Omega\,\rangle = \exp(-iE/\hbar)\bar{\Omega} \in \Phi$, then by definition this vector $|\bar{\Omega}\,\rangle \in \Phi$ enjoys the following eigenvalue equation: $\hat{H}|\bar{\Omega}\,\rangle = E|\bar{\Omega}\,\rangle$, (i.e. $(-\hbar/i)\partial/\partial t|\Omega\rangle = E|\Omega\rangle$!), where $E \in \mathbb{R}$ should be bounded from below, that is $E \;>> -\infty$. The latter means that the system is also stable. Due to the equation (5.3), one sees right away that

$$\begin{aligned}\frac{d}{dt}\langle\,\hat{N}\,\rangle &= \frac{d}{dt}\langle\,\Omega|\hat{N}|\Omega\rangle = \frac{d}{dt}\langle\,\bar{\Omega}|\hat{N}|\bar{\Omega}\,\rangle \\ &= \langle\,\bar{\Omega}|[\hat{H},\hat{N}]|\bar{\Omega}\,\rangle = \langle\,\bar{\Omega}|(\hat{H}\hat{N}-\hat{N}\hat{H})|\bar{\Omega}\,\rangle \\ &= \langle\,\bar{\Omega}|\hat{H}\hat{N}|\bar{\Omega}\,\rangle - \langle\,\bar{\Omega}|\hat{N}\hat{H}|\bar{\Omega}\,\rangle \\ &= \langle\,\bar{\Omega}|\hat{N}|\bar{\Omega}\,\rangle E - E\langle\,\bar{\Omega}|\hat{N}|\bar{\Omega}\,\rangle = 0 \end{aligned} \tag{5.4}$$

Thus, we found that the average particle number $N = \langle\,\hat{N}\,\rangle$ is constant at the equilibrium state. Subject to this state $|\Omega\rangle \in \Phi$ we can assume in general, that the Hamiltonian operator (5.1) is depending on time $t \in \mathbb{R}$: $\hat{H} = \hat{H}(t)$. Then for the evolution of any eigenstate $|\Omega_0\,\rangle \in \Phi$ one has the Schrödinger equation

$$\frac{d|\Omega(t)\,\rangle}{dt} = -\frac{i}{\hbar}\hat{H}(t)|\Omega(t)\,\rangle\,, \tag{5.5}$$

The usual way of solving (5.5) is the iterative procedure due to Picard:

$$|\Omega(t)\,\rangle = |\Omega(t_0)\,\rangle - \frac{i}{\hbar}\int_{t_0}^{t} dt_1\hat{H}(t_1)|\Omega(t_1)\,\rangle\,. \tag{5.6}$$

Having chosen $|\Omega(t_0)\,\rangle = |\Omega_0\,\rangle \in \Phi$, from (5.6) one follows that

$$\begin{aligned}|\Omega(t)\,\rangle = |\Omega_0\,\rangle &- \frac{i}{\hbar}\int_{t_0}^{t} dt_1\hat{H}(t_1)|\Omega_0\,\rangle \\ &+ \left(\frac{i}{\hbar}\right)^2\int_{t_0}^{t} dt_1\int_{t_0}^{t} dt_2\hat{H}(t_1)\hat{H}(t_2)|\Omega_0\,\rangle + \cdots \\ &+ \left(-\frac{i}{\hbar}\right)^n\int_{t_0}^{t} dt_1\int_{t_0}^{t} dt_2\cdots\hat{H}(t_n)|\Omega_0\,\rangle\end{aligned}$$

$$= \left(1 - \frac{i}{\hbar}\int_{t_0}^{t} dt_1 \hat{H}(t_1)\right.$$

$$\left. + \left(-\frac{i}{\hbar}\right)^2 \frac{1}{2!}\int_{t_0}^{t} dt_1 \int_{t_0}^{t} dt_2 T(\hat{H}(t_1)\hat{H}(t_2)) + \cdots \right) |\Omega_0\,\rangle$$

$$= T\exp\left(-\frac{i}{\hbar}\int_{t_0}^{t} \hat{H}(t')dt'\right) |\Omega_0\,\rangle = \hat{U}(t,t_0)|\Omega_0\,\rangle \qquad (5.7)$$

The operator T is the 'time ordering' operator while the operator $\hat{U}(t,t_0): \Phi \longrightarrow \Phi$ is called the 'evolution operator' of the system under consideration. From (5.7) one verifies easily that

$$\frac{d\hat{U}}{dt} = -\frac{i}{\hbar}\hat{H}(t)\hat{U}(t,t_0)\,, \qquad (5.8)$$

and $\hat{U}|_{t=t_o} = 1$ for all $t_0 \in \mathbb{R}$. Subject to the evolution (5.8) any operator $\hat{A}: \Phi \longrightarrow \Phi$ can be transformed into $\hat{A}(t): \Phi \longrightarrow \Phi$ via the rule

$$\hat{A}_H(t) = \hat{U}^+\hat{A}\hat{U} : \Phi \longrightarrow \Phi\,, \qquad (5.9)$$

where $\hat{U}^+\hat{U} = \hat{U}\hat{U}^+ = 1$. Then clearly,

$$\frac{d\hat{A}_H}{dt} = \frac{i}{\hbar}\hat{U}^+\hat{H}\hat{A}\hat{U} - \frac{i}{\hbar}\hat{U}^+\hat{A}\hat{H}\hat{U} = \frac{i}{\hbar}[\hat{H}_H(t), \hat{A}_H(t)]\,, \qquad (5.10)$$

where, also, $\hat{H}_H(t) = \hat{U}^+\hat{H}(t)\hat{U} = \hat{H}_H(0)$ due to (5.10) :

$$\frac{d\hat{H}_H}{dt} = \frac{i}{\hbar}[\hat{H}_H, \hat{A}_H] = 0 \quad \text{for all } t \in \mathbb{R}!\,.$$

The latter means apparently that $\hat{A}_H = const$ is $[\hat{H}_H, \hat{A}_H] = 0$ for all $t \in \mathbb{R}$.

In the case when $\hat{H} = \hat{H}_0 + \hat{H}_{int}(t)$, the vector of state $|\Omega\rangle \in \Phi$ can be represented a priori as

$$|\Omega\rangle = \exp\left(-\frac{i}{\hbar}\hat{H}_0 t\right)|\Omega_I\,\rangle\,, \qquad (5.11)$$

where, evidently due to (5.5) for $|\Omega_I\,\rangle \in \Phi$

$$\frac{d}{dt}|\Omega_I\,\rangle = -\frac{i}{\hbar}\hat{H}_I|\Omega_I\,\rangle,\ \hat{H}_I = \exp\left(\frac{i}{\hbar}t\hat{H}_0\right)\hat{H}_{int}\exp\left(-\frac{i}{\hbar}t\hat{H}_0\right), \qquad (5.12)$$

giving rise to the so called *interaction picture representation.* The solution to (5.12) (similarly to (5.5)) is given as

$$|\Omega_I\ \rangle = T\exp\left[-\frac{i}{\hbar}\int\limits_{t_0}^{t}\hat{H}_I(t')dt'\right]|\Omega_I(t_0)\ \rangle = \hat{U}_I(t,t_0)|\Omega_I(t_0)\ \rangle\,. \quad (5.13)$$

For the operator $\hat{U}_I(t,t_0)$ to be calculated more effectively, we consider the case when

$$\hat{H}_{int}(t) = \hat{H}_{int}\exp(-\varepsilon|t|) \quad (5.14)$$

as $\varepsilon\downarrow 0$. This means, that at $t=\pm\infty$ $\hat{H}(t)=\hat{H}_0$ as $\varepsilon\downarrow 0$, at $t=0$ $\hat{H}(t)=\hat{H}_0+\hat{H}_{int}$ as $\varepsilon\downarrow 0$, not depending on $t\in\mathbb{R}$ and $\varepsilon\downarrow 0$. So, if we are looking for the ground state vector $|\Omega\rangle\in\Phi$, (5.13) leads to

$$|\Omega\rangle = \exp\left(-\frac{i}{\hbar}\hat{H}_0\right)\hat{U}_{I,\varepsilon}(t)|\Omega_0\ \rangle|_{\varepsilon\downarrow 0,t=0} = \hat{U}_{I,\varepsilon}(0,-\infty)|\Omega_0\ \rangle|_{\varepsilon\downarrow 0}\,, \quad (5.15)$$

where, $|\Omega_0\ \rangle\in\Phi$ is the normalized ground state of the non-interacting part $\hat{H}_0$ of the Hamiltonian (5.1) by definition:

$$\hat{H}_0|\Omega_0(t_0)\ \rangle = E_0|\Omega_0\ \rangle,\ \langle\ \Omega_0|\Omega_0\ \rangle = 1\,. \quad (5.16)$$

Since the vector (5.15) is in general not normalized, it is necessary to define the following quasi-normalized expression before taking the limit $\varepsilon\downarrow 0$:

$$|\bar{\Omega}^{(\varepsilon)}\ \rangle = \frac{\hat{U}_I^{\varepsilon}(0,-\infty)|\Omega_0\ \rangle}{\langle\ \Omega_0|\hat{U}_I^{\varepsilon}(0,-\infty)|\Omega_0\ \rangle}\,, \quad (5.17)$$

where

$$\lim_{\varepsilon\downarrow 0}\langle\ \Omega_0|\bar{\Omega}^{(\varepsilon)}\ \rangle = 1\,. \quad (5.18)$$

Gell-Mann and Law proved that the vector (5.17) is really an eigenvector of the Hamiltonian (5.1), i.e.

$$\hat{H}|\bar{\Omega}^{(\varepsilon)}\ \rangle = E|\bar{\Omega}^{(\varepsilon)}\ \rangle\,. \quad (5.19)$$

But in general, the eigenvector $|\bar{\Omega}^{(\varepsilon)}\ \rangle\in\Phi$ should not necessarily be the ground state vector for (5.1); this is the state with the least energy!

This way now one can obtain the expression for the ground state energy $E \in \mathbb{R}^1$:

$$E - E_0 = \frac{\langle\, \Omega_0 | \hat{H}_I | \Omega \rangle}{\langle\, \Omega_0 | \Omega \rangle}\,, \tag{5.20}$$

where $\hat{H}_0 | \Omega_0\ \rangle = E_0 | \Omega_0\ \rangle$, $E_0 \in \mathbb{R}^1$ is the ground state energy of the $\hat{H}_0$-Hamiltonian.

Chapter 6

The One-Particle Green's Function

The one-particle Green's function (suppressing $\hbar$ for convenience in notation) is defined as

$$iG_{\alpha\beta}(\mathbf{x},t;\mathbf{x}',t') := \frac{\langle\, \Omega|T(\hat{\psi}_{\alpha,H}(\mathbf{x},t)\ \hat{\psi}^{+}_{\beta,H}(\mathbf{x}',t'))|\ \Omega\rangle}{\langle\, \Omega|\Omega\rangle}, \qquad (6.1)$$

where as before, the state $|\Omega\rangle \in \Phi$ is the ground state of the Hamiltonian (5.1), and need not necessarily be normalized here. If the Green's function (6.1) is found, one can obtain the following quantities: with Tr denoting the trace operator,

$$\langle\ \hat{\rho}(\mathbf{x})\ \rangle = \pm i Tr G(\mathbf{x},t;\mathbf{x},t^{+}) \qquad (6.2)$$

- the density of particles;

$$\langle\ \hat{\sigma}(\mathbf{x})\ \rangle = \pm i Tr(G(\mathbf{x},t;\mathbf{x},t^{+})\hat{\sigma}(\mathbf{x}) \qquad (6.3)$$

- the spin density;

$$\langle\ \hat{H}_0\ \rangle = \pm i \int\limits_{\mathbb{R}^3} d^3\mathbf{x} \lim_{\mathbf{x}'\to\mathbf{x}} \left[-\frac{\hbar^2\nabla^2_{\mathbf{x}}}{2m} G(\mathbf{x},t;\mathbf{x}',t^{+})\right] \qquad (6.4)$$

-the total kinetic energy;

$$\langle\ \hat{H}_{int}\ \rangle = \pm\frac{1}{2} i \int\limits_{\mathbb{R}^3} d^3\mathbf{x} \lim_{\mathbf{x}'\to\mathbf{x}} \left[Tr(i\hbar\frac{\partial}{\partial t} - \hat{T}\ (\mathbf{x}))G(\mathbf{x},t;\mathbf{x}',t^{+})\right] \qquad (6.5)$$

- the total interaction potential energy, and

$$E = \langle\ \hat{H}_0 + \hat{H}_{int}\ \rangle = \pm\frac{1}{2} i \int\limits_{\mathbb{R}^3} d^3\mathbf{x} \left(i\hbar\frac{\partial}{\partial t} - \frac{\hbar^2\nabla^2_{\mathbf{x}}}{2m}\right) Tr G(\mathbf{x},t;\mathbf{x}',t^{+}) \quad (6.6)$$

- the total ground state energy of the system, where by definition, $t^+ := \lim_{\varepsilon \downarrow 0}(t' + \varepsilon)$. All of the above results hold for both bosons and fermions.

Due to the uniformity of the space-time, the Green's function (6.10) depends only on the arguments $(t - t')$ and $(\mathbf{x} - \mathbf{x}')$. This motivates us to perform the Fourier transform of the expression (6.1)

$$G_{\alpha\beta}(\mathbf{x}, t; \mathbf{x}', t') = \sum_{\mathbf{k}} \int_{\mathbb{R}} \frac{d\omega}{2\pi} e^{i\langle\, \mathbf{k}, (\mathbf{x}-\mathbf{x}')\,\rangle - i\omega(t-t')} G_{\alpha\beta}(\mathbf{k}, \omega)\,, \tag{6.7}$$

that gives rise, in the limit $V \longrightarrow \infty$ to

$$G_{\alpha\beta}(\mathbf{x}, t; \mathbf{x}', t') = \frac{1}{(2\pi)^4} \int_{\mathbb{R}^3} d^3\mathbf{k} \int_{\mathbb{R}} d\omega \exp[i\langle\, \mathbf{k}, (\mathbf{x}-\mathbf{x}')\,\rangle - i\omega(t-t')] G_{\alpha\beta}(\mathbf{k}, \omega) \tag{6.8}$$

Thereby, for the total number of particles in the state $|\Omega\rangle$ (6.7) leads to

$$N = \int_V d^3\mathbf{x} \langle\, \hat{\rho}(\mathbf{x})\,\rangle = \pm i \frac{V}{(2\pi)^4} \lim_{\eta \downarrow 0} \int_{\mathbb{R}^3} d^3\mathbf{k} \int_{\mathbb{R}} d\omega e^{i\omega\eta} Tr G(\mathbf{k}; \omega)\,; \tag{6.9}$$

and for the total energy one arrives at

$$E = \pm \frac{i}{2} \frac{V}{(2\pi)^4} \lim_{\eta \downarrow 0} \int_{\mathbb{R}^3} d^3\mathbf{k} \int_{\mathbb{R}} d\omega e^{i\omega\eta} \left(\frac{\mathbf{k}^2 \hbar^2}{2m} + \hbar\omega \right) Tr G(\mathbf{k}; \omega)\,. \tag{6.10}$$

Chapter 7

The Ground States of Noninteracting Fermi- and Bose-Many Particle Systems

The ground state $|\Omega\rangle \in \Phi$ should obviously satisfy the following (eigenvalue) equation:

$$\hat{H}_0|\Omega_0\rangle = E_0|\Omega_0\rangle, \tag{7.1}$$

where $E_0 \quad >> -\infty$ is the ground state energy. Since $\hat{\psi}^+(\mathbf{x})$ and $\hat{\psi}(\mathbf{x}) : \Phi \longrightarrow \Phi$ are the standard creation and annihilation operators, their momentum representations $c^+_{\mathbf{k}}$ and $c_{\mathbf{k}} : \Phi \longrightarrow \Phi$ enjoy, the conditions $c_{\mathbf{k}}|0\rangle = 0$ ($[\hat{P}, \hat{H}_0] = 0$, $\hat{P} = \sum_{\mathbf{k}} \mathbf{k} c^+_{\mathbf{k}} c_{\mathbf{k}}$) for all $\mathbf{k} \in \mathbb{R}^3$. Thus, since our system consists of $N \in \mathbb{R}_+$ particles, the ground state should simultaneously enjoy the condition

$$\hat{N}|\Omega_0\rangle = N|\Omega_0\rangle. \tag{7.2}$$

Thereby for $|\Omega_0\rangle \in \Phi$ one can suggest the following expression for fermi particles:

$$|\Omega_0\rangle = \sum_{|\mathbf{k}| \le k_F} c^+_{\mathbf{k}}|0\rangle, \tag{7.3}$$

whence due to (7.2)

$$\sum_{|\mathbf{k}| \le k_F} 1 = N \simeq \frac{V}{(2\pi)^3} \int_{|\mathbf{k}| \le k_F} d^3\mathbf{k}, \tag{7.4}$$

where $k_F \quad > 0$ is called a Fermi sea parameter. As a result of (7.3) and (7.4)

$$E_0 = \langle \Omega_0| \sum_{|\mathbf{k}| \le k_F} \frac{\mathbf{k}^2}{2m} c^+_{\mathbf{k}} c_{\mathbf{k}} |\Omega_0\rangle = \frac{V}{(2\pi)^3} \int_{|\mathbf{k}| \le k_F} \frac{\mathbf{k}^2}{2m} d^3\mathbf{k} \tag{7.5}$$

The ground state $|\Omega_0\ \rangle \in \Phi$ represent a so called Fermi electron-hole sea being completely filled due to the absence of interaction between particles. Let us now consider the action of $(c_{\mathbf{k}}^{+}, c_{\mathbf{k}})-$ operators on the ground state $|\Omega_0\ \rangle$:

$$c_{\mathbf{k}}^{+}|\Omega_0\ \rangle = 0 \quad \text{if } |\mathbf{k}| \leq k_F\,, \tag{7.6}$$

and

$$c_{\mathbf{k}}|\Omega_0\ \rangle = 0 \quad \text{if } |\mathbf{k}|\ \rangle k_F\,. \tag{7.7}$$

Thus, we can define new creation annihilation operators $a_{\mathbf{k}}, b_{\mathbf{k}} : \Phi \longrightarrow \Phi$ as follows:

$$c_{\mathbf{k}} := \begin{cases} a_{\mathbf{k}}, & if\ |\mathbf{k}|\ \rangle k_F \\ b_{\mathbf{k}}^{+}, & if\ |\mathbf{k}| \leq k_F\,. \end{cases} \tag{7.8}$$

Similarly, their adjoint expressions are defined as follows:

$$c_{\mathbf{k}}^{+} := \begin{cases} a_{\mathbf{k}}^{+}, & if\ |\mathbf{k}|\ \rangle k_F \\ b_{-\mathbf{k}}, & if\ |\mathbf{k}| \leq k_F\,. \end{cases} \tag{7.9}$$

$|\mathbf{k}|\ > k_F$ defines 'particle' states $|\mathbf{k}| \leq k_F$, the 'hole' states. Figure 7.1 demonstrates schematically the created particle and hole states versus their energies. It is easy to see that the creation of both a particle and a hole *raises* the total energy of the system; so, $|\Omega_0\ \rangle \in \Phi$ is really the ground state!

Now it is easy calculate the free Green's function $G_{\alpha\beta}^{(0)}(\mathbf{x}, t; \mathbf{x}', t')$ or its Fourier-transform as

$$\begin{aligned} G_{\alpha\beta}^{(0)}(\mathbf{x}, t; \mathbf{x}', t') &= \delta_{\alpha\beta}\frac{1}{V}\sum_{\mathbf{k}} \exp(i\langle\ \mathbf{k}, (\mathbf{x}-\mathbf{x}')\ \rangle - i\omega(t-t')) \\ &\quad \times [\vartheta(t-t')\vartheta(\mathbf{k}-k_F) - \vartheta(t-t')\vartheta(k_F-\mathbf{k})] \\ &= \frac{\delta_{\alpha\beta}}{(2\pi)^4}\int\limits_{\mathbb{R}^3} d^3\mathbf{k} \int\limits_{\mathbb{R}^3} d\omega \exp(i\langle\ \mathbf{k}, (\mathbf{x}-\mathbf{x}')\ \rangle - i\omega(t-t')) \\ &\quad \times \left[\frac{\vartheta(\mathbf{k}-k_F)}{\omega-\omega_{\mathbf{k}}+i\eta} + \frac{\vartheta(k_F-\mathbf{k})}{\omega-\omega_{\mathbf{k}}+i\eta}\right], \end{aligned} \tag{7.10}$$

or

$$G_{\alpha\beta}^{(0)}(\mathbf{k};\omega) = \delta_{\alpha\beta}\left[\frac{\vartheta(\mathbf{k}-k_F)}{\omega-\omega_{\mathbf{k}}+i\eta} + \frac{\vartheta(k_F-\mathbf{k})}{\omega-\omega_{\mathbf{k}}+i\eta}\right], \tag{7.11}$$

where we made use of the identity

$$\vartheta(t-t') = -\frac{1}{2\pi i}\int\limits_{\mathbb{R}^1} \frac{e^{i\omega(t-t')}}{\omega+i\eta}d\omega \tag{7.12}$$

at $\eta \downarrow 0$. The corresponding noninteracting Hamiltonian $\hat{H}_0$ (restoring $\hbar$ into the equation) takes the form:

$$\hat{H}_0 = \sum_{\alpha,|\mathbf{k}|\ \rangle k_F} \hbar\omega_{\mathbf{k}} a^+_{\mathbf{k}\alpha} a_{\mathbf{k}\alpha} - \sum_{\alpha,|\mathbf{k}|\le k_F} \hbar\omega_{\mathbf{k}} a^+_{\mathbf{k}\alpha} b_{\mathbf{k}\alpha} + \sum_{\alpha,|\mathbf{k}|\le k_F} \hbar\omega_{\mathbf{k}}\,, \tag{7.13}$$

that is, the energy operator consists of positive energy operator of particles and negative energy operator of holes plus the total energy of the filled Fermi sea E_0 found before. As usual, based on (7.8) and (7.9), one finds that

$$[a_{\mathbf{s}}, a^+_{\mathbf{k}}]_+ = \delta_{\mathbf{ks}} = [b_{\mathbf{k}}, b^+_{\mathbf{s}}]_+\,, \tag{7.14}$$

is the standard fermi-statistics relationships.

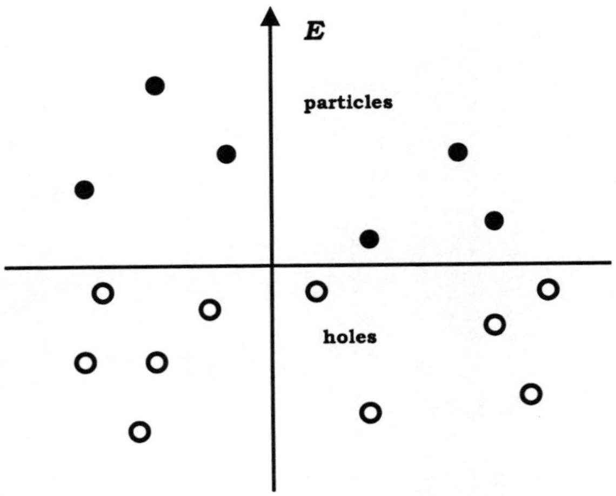

Fig. 7.1.

Chapter 8

The Wick Theorem and Feynman Diagrammatic Expansion

We are now in a position to calculate the Green's function (6.3) via the Feynman perturbation technique. Let's take any operator $\hat{A}_I(t)$ and $\hat{B}_I(t)$ in the interaction picture. We are interested in calculating the average value of

$$\frac{\langle\, \Omega | T(\hat{A}_H(t)\hat{B}_H(t'))|\Omega\,\rangle}{\langle\, \Omega|\Omega\rangle} = \overbrace{\hat{A}(t)\hat{B}(t')}\,, \tag{8.1}$$

where $|\Omega\rangle \in \Phi$ is the ground eigenvalue state of our Hamiltonian (5.1). Owing to the result obtained before, we can write down:

$$|\Omega\rangle = \frac{\hat{U}_I^\varepsilon(0;\pm\infty)|\Omega_0\,\rangle}{\langle\, \Omega_0|\hat{U}_I^\varepsilon(0;\pm\infty)|\Omega_0\,\rangle}\,, \tag{8.2}$$

as $\varepsilon \downarrow 0$ and substitute it into (8.1). Then

$$\overbrace{\hat{A}(t)\hat{B}(t')} = \frac{\langle\, \Omega_0|\hat{U}_I^\varepsilon(\infty,0)\hat{U}_I^\varepsilon(0;t)\hat{A}_I(t)\hat{U}_I^\varepsilon(t,t')\hat{B}_I(t')\hat{U}_I^\varepsilon(t',-\infty)|\Omega_0\,\rangle}{\langle\, \Omega_0|\hat{U}_I^\varepsilon(\infty,-\infty)|\Omega_0\,\rangle}\Big|_{\varepsilon\downarrow 0}$$

$$= \frac{\langle\, \Omega_0|T(\hat{S}_I^\varepsilon\hat{A}_I(t)\hat{B}_I(t'))|\Omega_0\,\rangle}{\langle\, \Omega_0|\hat{S}_I^\varepsilon|\Omega_0\,\rangle}\Bigg|_{\varepsilon\downarrow 0}\,, \tag{8.3}$$

where by definition, $\hat{S}_I^\varepsilon := \hat{U}_I^\varepsilon(+\infty;-\infty)$ is the so called scattering operator. Recalling the definition of the $\hat{U}_I^\varepsilon(t,t')-$ evolution operator, (8.3) results in

$$\overbrace{\hat{A}(t)\hat{B}(t')} = \sum_{n=0}^{\infty}\left(-\frac{i}{\hbar}\right)^n \int\limits_{-\infty}^{\infty} dt_1 \cdots \int\limits_{-\infty}^{\infty} dt_n$$

$$\times \langle\, \Omega_0|T(\hat{H}_I(t_1)\hat{H}_I(t_2)\cdots\hat{H}_I(t_n)\hat{A}_I(t)\hat{B}_I(t'))|\Omega_0\,\rangle\,. \tag{8.4}$$

In the case when operator $[\hat{N},\hat{H}]=0$, the interaction operator $\hat{H}_{int}$ should contain an equal amount of creation and annihilation operators and can generally be expressed as

$$\begin{aligned}\hat{H}_{int} &= \frac{1}{2}\sum_{\substack{\mathbf{k},\mathbf{s},\mathbf{q}\\ \alpha,\alpha';\beta,\beta'}} U^{(2)}(\mathbf{q})_{\alpha\alpha';\beta\beta'} c^{+}_{\mathbf{k},\alpha} c^{+}_{\mathbf{s},\beta} c_{\mathbf{s}+\mathbf{q},\beta'} c_{\mathbf{k}-\mathbf{q},\alpha'} \\ &= \frac{1}{2}\sum_{\alpha,\alpha';\beta,\beta'} \int_{\mathbb{R}^4} d^4\mathbf{x}\cdots\int_{\mathbb{R}^4} d^4\mathbf{x} \\ &\quad \times U^{(2)}(\mathbf{x},\mathbf{x}')_{\alpha\alpha';\beta\beta'}\hat{\psi}^{+}_{\alpha}(\mathbf{x})\hat{\psi}^{+}_{\beta}(\mathbf{x}')\hat{\psi}_{\beta'}(\mathbf{x}')\hat{\psi}_{\alpha'}(\mathbf{x}),\end{aligned} \tag{8.5}$$

where, by definition, $X=(\mathbf{x},t)$, $X'=(\mathbf{x}',t')\in\mathbb{R}^4$. Taking now $\hat{A}=\hat{\psi}_\alpha$ and $\hat{B}=\hat{\psi}^{+}_{\beta}$–operators, we find that

$$\begin{aligned}&iG_{\alpha\beta}(X;X') \\ &= iG^{(0)}_{\alpha\beta}(X;X') \\ &\quad+\left(-\frac{i}{\hbar}\right)\int_{-\infty}^{\infty} dt_1\langle\,\Omega_0|T(\hat{H}_I(t_1)\hat{\psi}_\alpha(X)\hat{\psi}^{+}_{\beta}(X'))|\Omega_0\,\rangle \\ &\quad+\left(-\frac{i}{\hbar}\right)^2\int_{-\infty}^{\infty} dt_1\int_{-\infty}^{\infty} dt_2\langle\,\Omega_0|T(\hat{H}_I(t_1)\hat{H}_I(t_2)\hat{\psi}_\alpha(X)\hat{\psi}^{+}_{\beta}(X'))|\Omega_0\,\rangle \\ &\quad+\cdots \\ &= iG^{(0)}_{\alpha\beta}(X;X')+\left(-\frac{i}{\hbar}\right)\sum_{\lambda,\lambda';\mu,\mu'}\int_{\mathbb{R}^4} d^4X_1\int_{\mathbb{R}^4} d^4X'_1 U^{(2)}_{\lambda,\lambda';\mu,\mu'}(X_1,X'_1) \\ &\quad\times\langle\,\Omega_0|T(\hat{\psi}^{+}_{\lambda}(X_1)\hat{\psi}^{+}_{\mu}(X'_1)\hat{\psi}_{\mu'}(X'_1)\hat{\psi}_{\lambda'}(X_1)\hat{\psi}_\alpha(X)\hat{\psi}^{+}_{\beta}(X'))|\Omega_0\rangle+\cdots.\end{aligned} \tag{8.6}$$

To calculate the expression (8.6) explicitly, it is useful to make use of the Wick theorem:

$$\begin{aligned}T(UVW\cdots XYZ) &= N(UVW\cdots XYZ)+N(\overbrace{UV}W\cdots XYZ) \\ &\quad+N(\overbrace{UVW}\cdots XYZ)+N(\overbrace{UV}\,\overbrace{W\cdots}\,\overbrace{\cdots X}\,\overbrace{YZ}),\end{aligned} \tag{8.7}$$

where, by definition,

$$T(UV) = N(UV) + \overbrace{U\,V} \tag{8.8}$$

with $N(\cdots)$ denoting the standard normal ordering. Owing to the fact that $N(\cdots)|\Omega_0\,\rangle = 0$, the expression (8.6) can be represented only by means of contractions $\overbrace{\hat{\psi}_\alpha(X)\ \hat{\psi}^+_\beta(Y)}$ of operator quantities in the interaction representation under consideration. Consequently, the expression (8.6) takes the form:

$$iG_{\alpha\beta}(X,Y) = \sum_{m\geq 0}\left(-\frac{i}{\hbar}\right)^m \frac{1}{m!}\int_{\mathbb{R}} dt_1 \cdots \int_{\mathbb{R}} dt_2$$
$$\times \langle\, \Omega_0|T\ (\hat{\psi}_\alpha(X)\hat{\psi}^+_\beta(Y)[\hat{H}_I(t_1)\hat{H}_I(t_2)\cdots \hat{H}_I(t_m)])|\Omega_0\,\rangle_{connected}\,, \tag{8.9}$$

or in the diagrammatic form due to Feynman diagrams

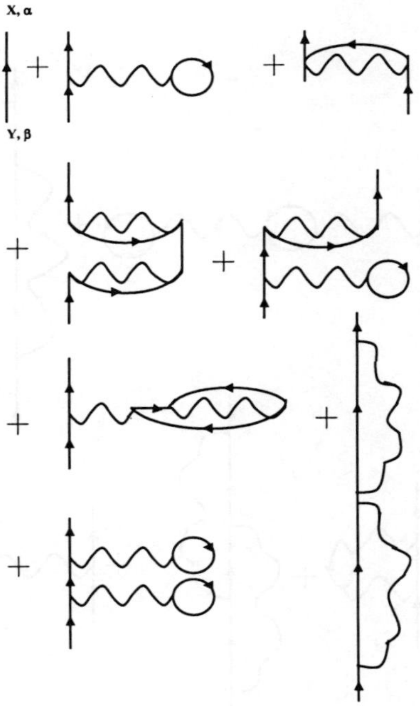

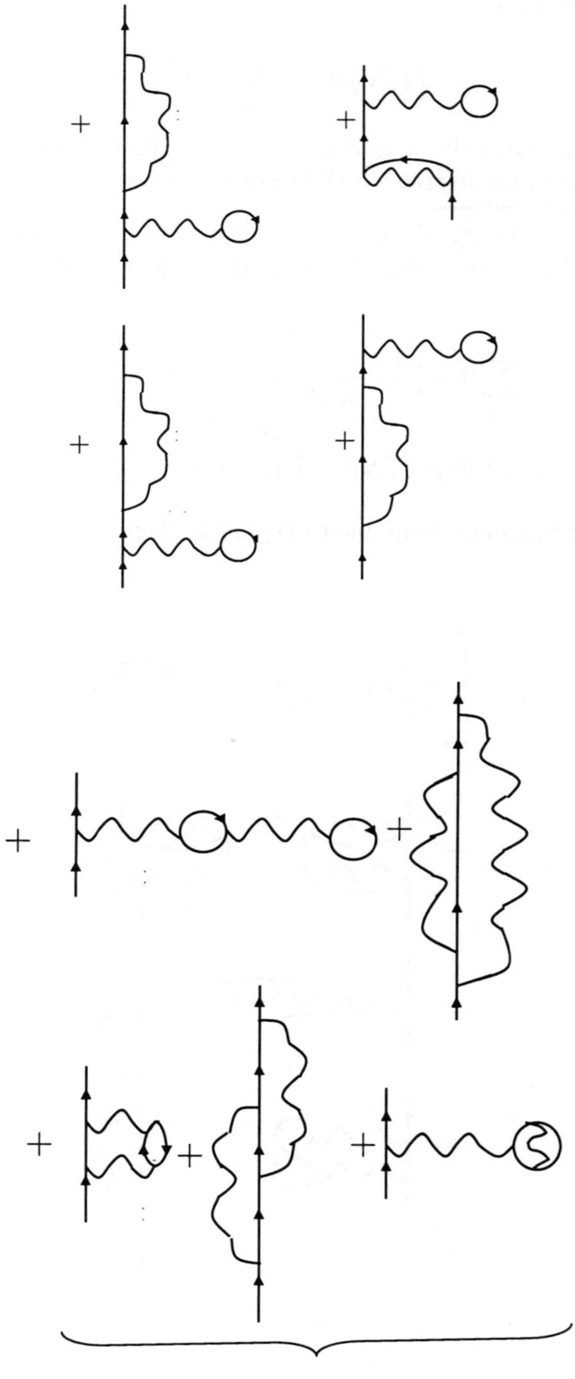

All second order diagrams

In the Fourier transformed form the analytical expression assigned to each diagram is simplified considerably. For instance, the first order diagrams are drawn as follows:

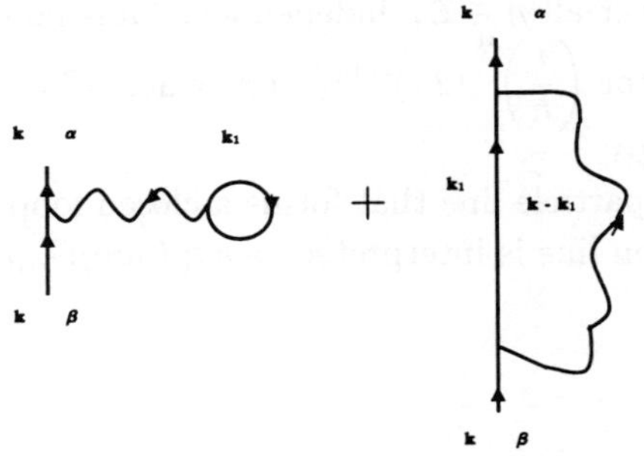

Fig. 8.3.

and the corresponding contribution to $G^{(1)}_{\alpha\beta}(k)$ is equal to the expression:

$$\begin{aligned} G^{(1)}_{\alpha\beta}(k) &= \left(\frac{i}{\hbar}\right)(-1)\frac{1}{(2\pi)^4}\int \mathbf{d}^4\mathbf{k}_1 G^{(0)}_{\alpha\lambda}(k)U^{(2)}(0)_{\lambda'\lambda;\mu\mu'} \\ &\times G^{(0)}_{\lambda'\beta}(k)G^{(0)}_{\mu\mu'}(k_1)e^{i\omega_1\eta} \\ &+ \left(\frac{i}{\hbar}\right)\frac{1}{(2\pi)^4}\int \mathbf{d}^4\mathbf{k}_1 G^{(0)}_{\alpha\lambda}(k)U^{(2)}(\mathbf{k}-\mathbf{k}_1)_{\lambda'\lambda;\mu\mu'} \\ &\times G^{(0)}_{\lambda'\mu}(k_1)G^{(0)}_{\mu'\beta}(k)e^{i\omega_1\eta}\,, \end{aligned} \tag{8.10}$$

where $k = (\mathbf{k}, \omega)$ and the spin summation has been simplified with the Kroenecker delta for each factor $G^{(0)}(k)$.

Feynman's rules: for n'th order:

i) Draw all topologically *distinct* connected diagrams with $n \in \mathbb{Z}_+$ interaction lines and $(2n+1)$ directed Green's functions of 0'th order;

ii) Assign a direction to each interaction line; associate a directed four-momentum with each line and *conserve* four-momentum at each vertex;

iii) Each Green's function corresponds to a factor

$$\begin{aligned} G^{(0)}_{\alpha\beta}(k) &= G^{(0)}_{\alpha\beta}(\mathbf{k},\omega) \Longrightarrow \delta_{\alpha\beta}G^{(0)}(\mathbf{k},\omega) \\ &= \delta_{\alpha\beta}\left[\frac{\vartheta(|\mathbf{k}|-k_F)}{\omega-\omega_{\mathbf{k}}+i\eta} + \frac{\vartheta(k_F-|\mathbf{k}|)}{\omega-\omega_{\mathbf{k}}+i\eta}\right] \end{aligned} \tag{8.11}$$

iv) Each interaction corresponds to a factor $U^{(2)}(k)_{\lambda'\lambda;\mu\mu'}$, where indices are associated to fermion lines;

v) Perform a spin summation along each continuous particle line including the potential at each vertex;

vi) Integrate over all $n \in \mathbb{Z}_+$ independent four-momenta;

vii) Affix a factor $\left(\frac{i}{\hbar}\right)^n (2\pi)^{-4n}(-1)^C$ where $C \in \mathbb{Z}_+$ is the number of closed fermion loops;

ix) Any single-particle line that forms a closed loop or that is linked by the same interaction line is interpreted as $\exp(i\omega\eta)G_{\alpha\beta}(k;\omega)$, where $\eta \to 0$ at the end.

Chapter 9

The Dyson Equations

If all connected diagrams are gathered together into one expression $\sum$, one evidently finds the following diagram expression:

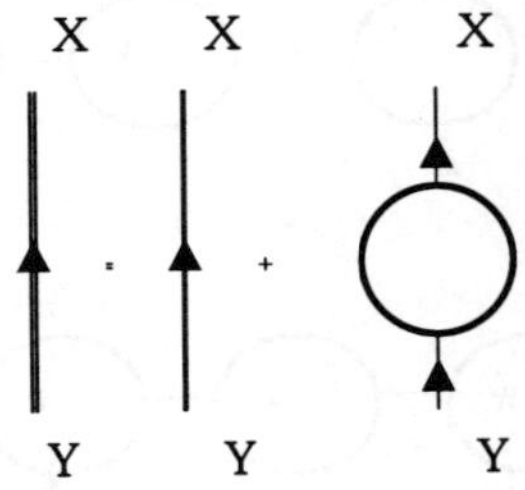

(9.1)

This insertion Σ is called a "self energy" part and satisfies the following analytical expression due to (9.1):

$$G_{\alpha\beta}(X,Y) = G^{(0)}_{\alpha\beta}(X,Y) + \int d^4X_1 \int d^4X_1' G^{(0)}_{\alpha\lambda}(X,X_1) \sum_{\lambda\mu}(X_1,X_1') G^{(0)}_{\mu\beta}(X_1',Y) . \quad (9.2)$$

It is useful to define a so called *proper* self-energy insertion $\sum^*$ as follows: *it is any part of a diagram that* is connected to the rest of the diagram by two particle lines and *can not* be separated *into two pieces* by *cutting a single particle-line.* By definition, the proper self-energy term is the sum of all proper self-energy insertions. The equation (9.2) can be iterated giving rise to the following expression:

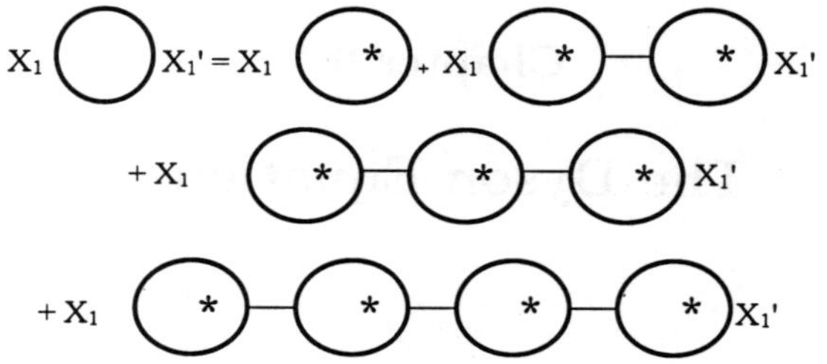

(9.3)

As a result of (9.3), equation (9.1) can be rewritten in the following form:

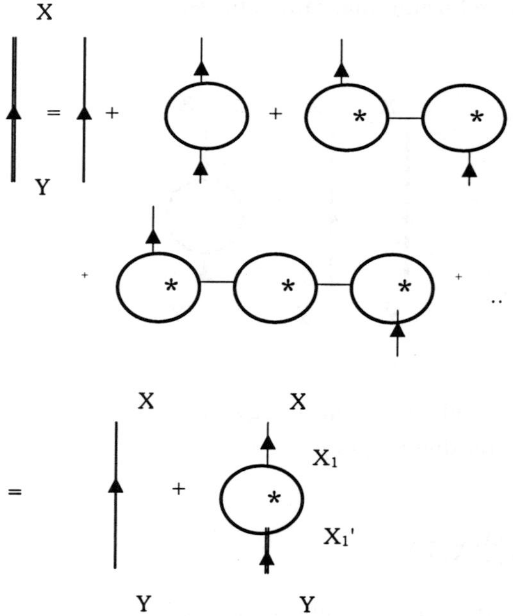

(9.4)

or in the analytical form

$$G_{\alpha\beta}(X,Y) = G^{(0)}_{\alpha\beta}(X,Y) +$$

$$\int d^4X_1 \int d^4X_1' G^{(0)}_{\alpha\gamma}(X,X_1) \sum_{\gamma\mu}(X_1,X_1') G_{\mu\beta}(X_1',Y) \quad (9.5)$$

Dyson equation becomes naturally much more simpler if the interaction is invariant under transformation and the system is spatially uniform. Then it is possible to introduce Fourier transforms to (9.2)

$$G_{\alpha\beta}(k) = G^{(0)}_{\alpha\beta}(k) + G^{(0)}_{\alpha\lambda}(k) \sum_{\lambda\mu}(k) G^{(0)}_{\mu\beta}(k) \tag{9.6}$$

with no integration over momentum variables! Similarly for (9.5) one gets *the algebraic equation:*

$$G_{\alpha\beta}(k) = G^{(0)}_{\alpha\beta}(k) + G^{(0)}_{\alpha\lambda}(k)\Sigma^*_{\lambda\mu}(k)G_{\mu\beta}(k) \tag{9.7}$$

Based on analytical study $\omega \in \mathbb{C}$ properties of the Green function (9.7), and due to the Lehmann theorem, one can state that:

i) singularities of the Green's function $G_{\alpha\beta}(\mathbf{k}, \omega)$ considered as a function of $\omega \in \mathbb{C}$, determine both the excitation energies $\varepsilon_{\mathbf{k}}$ of the system and their dampings $\gamma_{\mathbf{k}}$;

ii) for real $\omega \in \mathbb{R}$ one has the following expressions:

$$\begin{aligned} \operatorname{Im}\Sigma^*(\mathbf{k};\omega) &\geq 0 \text{ if } \hbar\omega < \mu\,, \\ \operatorname{Im}\Sigma^*(\mathbf{k};\omega) &\leq 0 \text{ if } \hbar\omega > \mu\,, \end{aligned} \tag{9.8}$$

meaning that the chemical potential of the system (interacting!) is defined by the point $\mu \in \mathbb{R}$ where $\operatorname{Im}\Sigma^*(\mathbf{k};\omega)$ changes its sign!

Consider the case when there is no spin interaction between particles. Then equation (9.7) can be solved exactly:

$$G_{\alpha\beta}(k) = G_{\alpha\beta}(\mathbf{k}, \omega) = \frac{1}{\dfrac{\omega - \varepsilon^0_{\mathbf{k}}}{\hbar} - \sum^*(\mathbf{k}, \omega)} \delta_{\alpha\beta}\,. \tag{9.9}$$

Subject to calculation of all components of the proper self-energy insertion Σ^*, we evidently have the expansion:

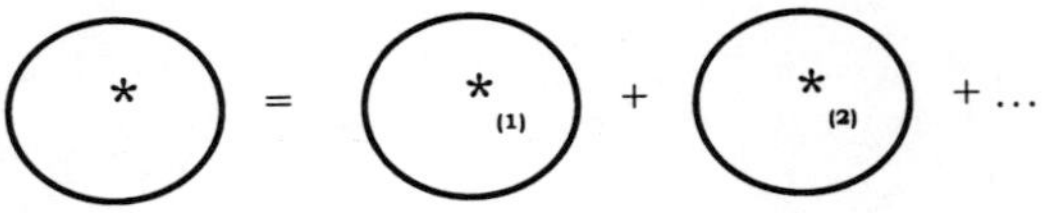

(9.10)

where, by definition,

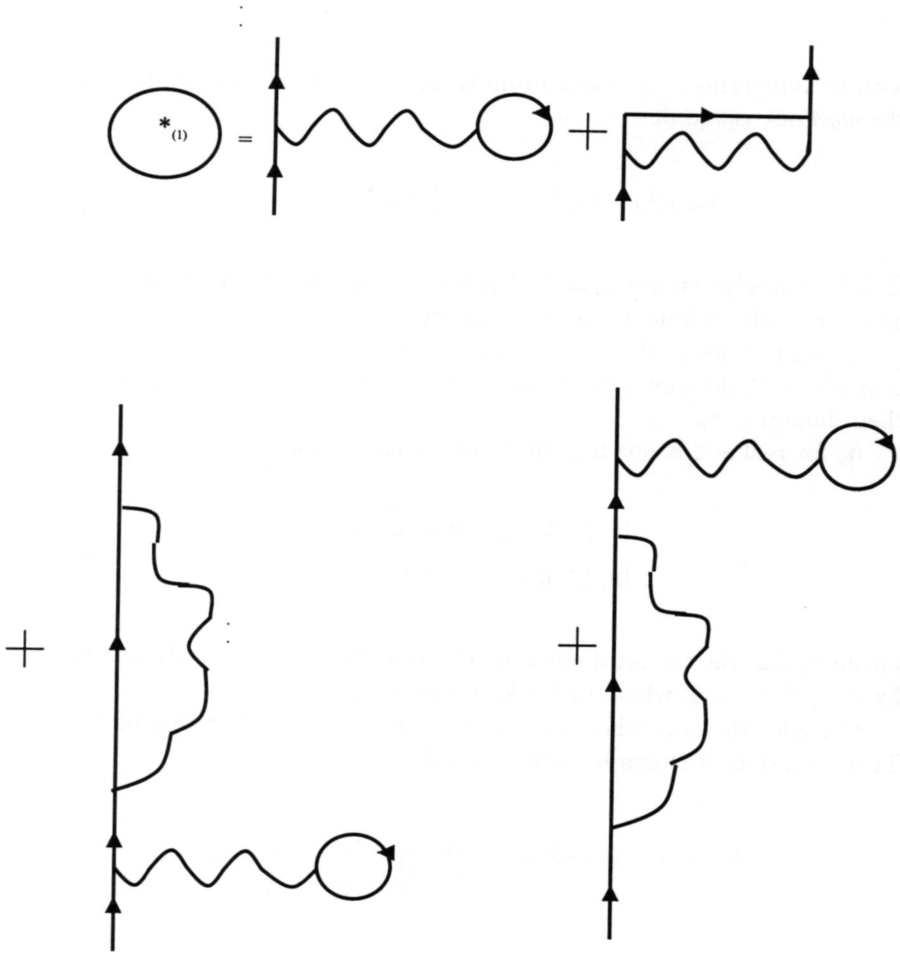

(9.11)

- all first-order terms,

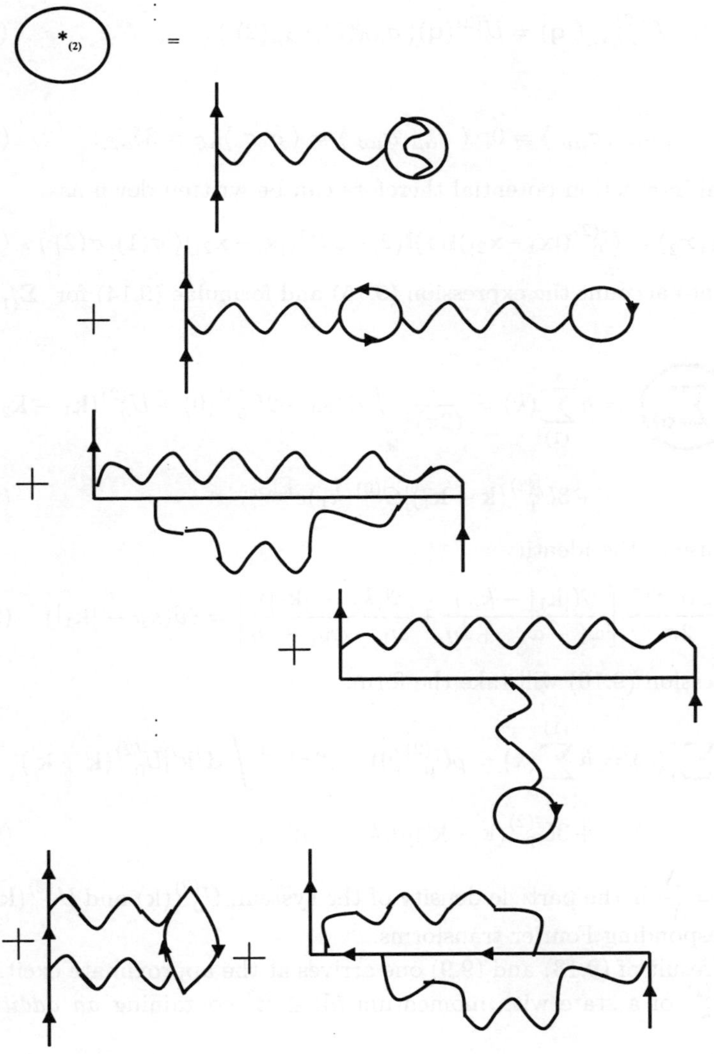

(9.12)

- all second-order terms of the self-energy.

We introduce here some remarks concerning interaction potentials between particles depending on spin; two most commonly used potentials in the literature are:

$$i) \qquad U^{(2)}_{\alpha\beta,\lambda\mu} = U^{(2)}(\mathbf{q})\delta_{\alpha\beta}\delta_{\gamma\mu} \sim \; U^{(2)} = U^{(2)}(\mathbf{q})\mathbb{I}(1)\mathbb{I}(2);$$

$$ii) \qquad U^{(2)}_{\alpha\beta,\lambda\mu}(\,\mathbf{q}) = U^{(2)}(\mathbf{q})\langle\, \sigma_{\alpha\beta}(1), \sigma_{\lambda\mu}(2)\,\rangle\,, \tag{9.13}$$

where

$$\langle\, \sigma_{\alpha\beta}, \sigma_{\mu\mu}\,\rangle = 0, \;\langle\, \sigma_{\alpha\mu}, \sigma_{\mu\beta}\,\rangle = \langle\, \sigma, \sigma\,\rangle_{\alpha\beta} = 3\delta_{\alpha\beta}\,. \tag{9.14}$$

The total interaction potential therefore can be written down as

$$U^{(2)}(\mathbf{x}_1, \mathbf{x}_2) = U^{(2)}_0(|\mathbf{x}_1 - \mathbf{x}_2|)\mathbb{I}(1)\mathbb{I}(2) + U^{(2)}_1(|\mathbf{x}_1 - \mathbf{x}_2|)\langle\, \sigma(1), \sigma(2)\,\rangle\,. \tag{9.15}$$

Taking into account the expression (9.15) and formulae (9.14) for $\Sigma^*_{(1)}$ one gets:

$$\hbar \bigcirc\!\!\!\!\!\!\!\!\Sigma^{*}_{(1)} = \hbar \sum_{(1)}^{*}(k) = \frac{i}{(2\pi)^4}\int_{\mathbb{R}^4} \mathbf{d}^4\mathbf{k}_1[-2U^{(2)}_0(0) + U^{(2)}_0(\mathbf{k}_1 - \mathbf{k}_2)$$

$$+ 3U^{(2)}_1(\mathbf{k} - \mathbf{k}_1)]G^{(0)}(k_1)e^{i\omega_1\eta}\,. \tag{9.16}$$

Making use of the identity

$$\int_{-\infty}^{\infty} \frac{d\omega_1 e^{i\omega_1\eta}}{2\pi}\left[\frac{\vartheta(|\mathbf{k}_1| - k_F)}{\omega_1 - \omega_{\mathbf{k}_1} + i\eta} + \frac{\vartheta(k_F - |\mathbf{k}_1|)}{\omega_1 - \omega_{\mathbf{k}_1} - i\eta}\right] = i\vartheta(k_F - |\mathbf{k}_1|)\,, \tag{9.17}$$

the expression (9.16) will take the form:

$$\hbar \sum\nolimits_1^{*}(k) = \hbar \sum^{(1)}(k) = \rho U^{(2)}_0(0) - (2\pi)^{-3}\int \mathbf{d}^3\mathbf{k}'[U^{(2)}_0(\mathbf{k} - \mathbf{k}')$$

$$+ 3U^{(2)}_1(\mathbf{k} - \mathbf{k}')]\vartheta(k_F - |\mathbf{k}'|)\,, \tag{9.18}$$

where $\rho = \dfrac{N}{V}$ is the particle density of the system, $U^{(2)}_0(\mathbf{k})$ and $U^{(2)}_1(\mathbf{k})$ are the corresponding Fourier transforms.

As a result of (9.18) and (9.9) one arrives at the approximate excitation energy $\varepsilon^{(1)}_{\mathbf{k}}$ of a state with momentum $\hbar\mathbf{k} \in \mathbb{R}^3$ containing *an additional* particle:

$$\varepsilon^{(1)}_{\mathbf{k}} = \varepsilon^{(0)}_{\mathbf{k}} + \hbar \sum\nolimits^{*}_{(1)}(\mathbf{k}) = \frac{\hbar^2|\mathbf{k}|^2}{2m} + \rho U^{(2)}_0(0)$$

$$- \frac{1}{(2\pi)^3}\int d^3\mathbf{k}'[U^{(2)}_0(\mathbf{k} - \mathbf{k}') + 3U^{(2)}_1(\mathbf{k} - \mathbf{k}')]\vartheta(k_F - |\mathbf{k}'|) \tag{9.19}$$

Based on the Green's function expression (9.9), one can now derive the explicit ground-state energy of a uniform system of particles (s-spin fermion):

$$\begin{aligned} E &= -iV(2s+1)\int \frac{d^4k}{(2\pi)^4} e^{i\omega\eta}\frac{\hbar}{2}\left[\frac{\hbar\omega+\varepsilon^0_{\mathbf{k}}}{\hbar\omega-\varepsilon^0_{\mathbf{k}}-\hbar\sum^*(\mathbf{k};\omega)}\right] \\ &= -iV(2s+1)\int \frac{d^4k}{(2\pi)^4} e^{i\omega\eta}\left\{\left[\varepsilon^0_{\mathbf{k}}-\frac{\hbar}{2}\Sigma^*(\mathbf{k};\omega)\right]G(\mathbf{k};\omega)+\frac{\hbar}{2}\right\} \\ &\Longrightarrow -iV(2s+1)(2\pi)^{-4}\int d^4k e^{i\omega\eta}\left\{\left[\varepsilon^0_{\mathbf{k}}+\frac{\hbar}{2}\sum^*(\mathbf{k};\omega)\right]G(\mathbf{k};\omega)\right]\right\}. \end{aligned} \tag{9.20}$$

Chapter 10

Polarization Operators

If one considers an inter-particle interaction potential $U^{(2)}$ like (9.15), the effective inter-particle interaction potential $U^{(2)*}$ will evidently differ from $U^{(2)}$due to the polarization of the ground state $|\Omega_0\ \rangle \in \Phi$ caused by the inter-particle interaction between ground state particles. To find this alteration let us apply the Feynman diagram approach developed above. We get:

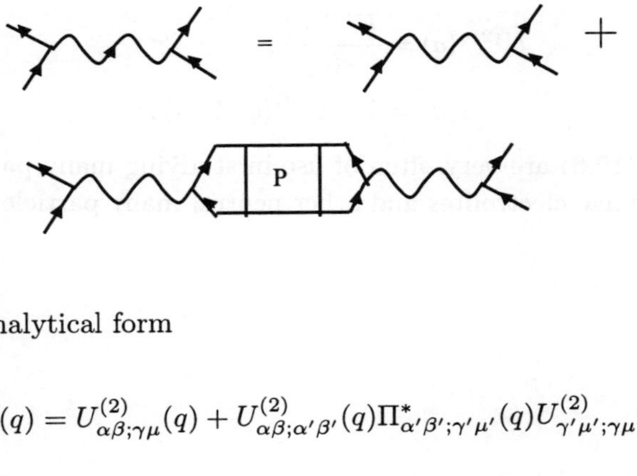

(10.1)

or in the analytical form

$$U^{(2)*}_{\alpha\beta;\gamma\mu}(q) = U^{(2)}_{\alpha\beta;\gamma\mu}(q) + U^{(2)}_{\alpha\beta;\alpha'\beta'}(q)\Pi^{*}_{\alpha'\beta';\gamma'\mu'}(q)U^{(2)}_{\gamma'\mu';\gamma\mu}(q)\,, \tag{10.2}$$

where $\Pi(q)$, $q = (\mathbf{q},\omega) \in \mathbb{R}^4$, is called a *polarization insertion.* As before, it is convenient to introduce the corresponding *proper polarization insertion* $\Pi^*(q)$, which is a *polarazion part* that can not be separated into two polarization parts by *cutting a single interaction line.* As a result of iteration applied to (10.2) one arrives at the following equivalent expression:

$$U^{(2)*}_{\alpha\beta;\gamma\mu}(q) = U^{(2)}_{\alpha\beta;\gamma\mu}(q) + U^{(2)}_{\alpha\beta;\beta'\alpha'}(q)\Pi^{*}_{\alpha'\beta';\gamma'\mu'}(q)U^{(2)*}_{\gamma'\mu';\gamma\mu}(q)\,, \tag{10.3}$$

or the same in the diagram form as

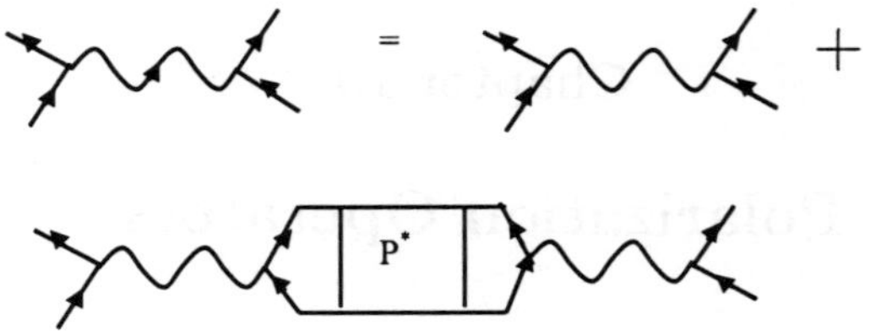

(10.4)

This is known as a Dyson equation for the proper polarization insertion Π^*.

In the case when there is no spin interaction between particles, from (10.4) one easily finds that

$$\begin{aligned} U^{(2)*}_{\alpha\beta;\gamma\mu}(q) &= U^{(2)*}(q)\delta_{\alpha\beta}\delta_{\gamma\mu}\,, \\ U^{(2)*}(q) &= \frac{U^{(2)}(q)}{1 - \Pi^*(q)U^{(2)*}(q)}\,, \end{aligned} \tag{10.5}$$

that is we obtained a so called screening potential $U^*(q)$ is obtained, defining a generalized dielectric function $\varepsilon(q)$ as follows:

$$\begin{aligned} U^{(2)*}(q) &= \frac{U^{(2)}(q)}{\varepsilon(q)}\,, \\ \varepsilon(q) &= 1 - U^{(2)}(q)\Pi^*(q)\,. \end{aligned} \tag{10.6}$$

Formulas like (10.6) are very often of use in studying many particle interaction systems like electrolites and other neutral many particle systems.

Chapter 11

The Hartree-Fock Approximation

As is well known from experiments, a single-particle description *forms* a surprisingly good approximation in many systems like metals, atoms, nuclei. Hence, a *natural* approach is to retain the single-particle picture and assume that "each particle moves in a single-particle potential (effective!) that comes from its average interaction with all the other particles". The single-particle energy should then be the *unperturbed* energy plus the potential energy of the interaction *averaged* over the *states occupied by all the other* particles.

We start from the proper self-energy insertion of first order:

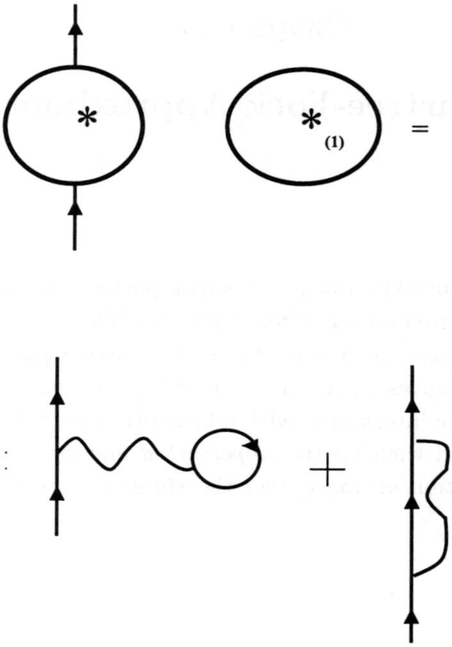

(11.1)

Based on the reasoning above, one can introduce the proper self-energy insertion into each Green's function line in (11.1), giving rise to the expression:

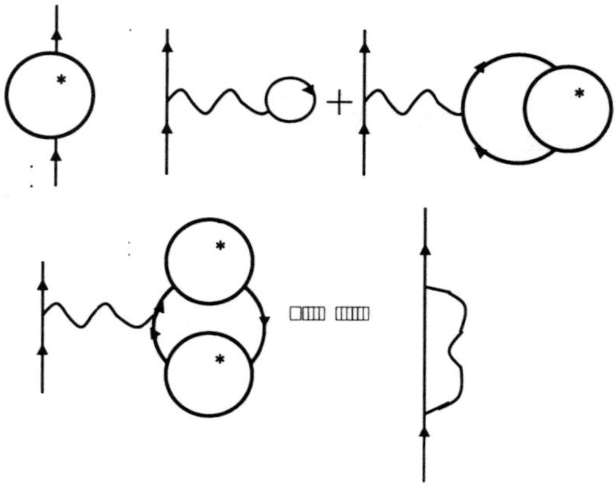

(11.2)

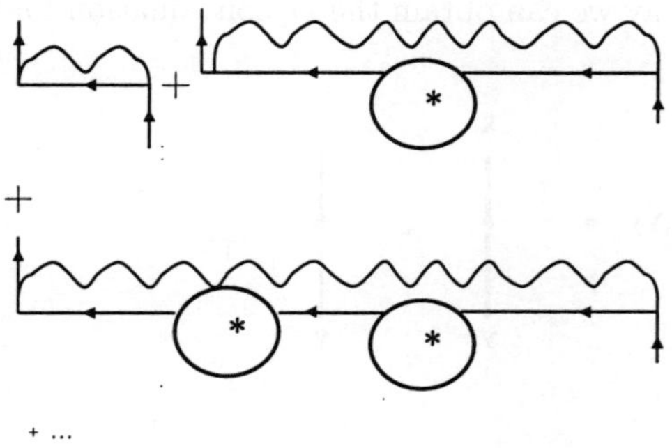

Here, we have obtained the series for proper self-energy Σ^* in *the Hartree - Fock* approximation. The diagrams (11.2) can be summed up to the following short form:

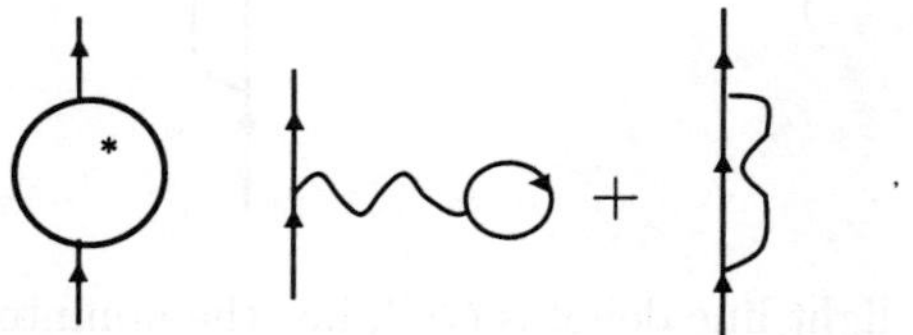

(11.3)

where, as before, the diagrams

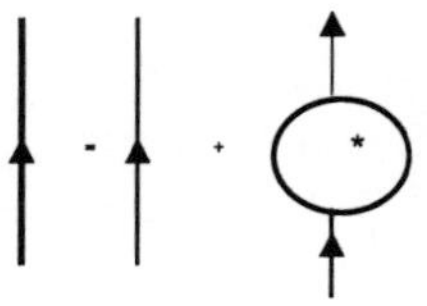

(11.4)

determine the Dyson equation for Green's function $G(X, Y)$. Subject to the insertion $\sum^*(X_1, X_1')$, we have the following exact expression:

$$\hbar \sum\nolimits^* (\mathbf{x}_{1,} t_{1;} \mathbf{x}_1', t_1')$$

$$= -i\delta(t_1 - t_1')[\delta(\mathbf{x}_1 - \mathbf{x}_1')(2s+1) \int d^3\mathbf{x}_2 G(\mathbf{x}_2 t_2; \mathbf{x}_2 t_2^+)$$

$$\times U^{(2)}(\mathbf{x}_1 - \mathbf{x}_2) - U^{(2)}(\mathbf{x}_1 - \mathbf{x}_1') G(\mathbf{x}_1 t_1; \mathbf{x}_1' t_1^+)] \, , \qquad (11.5)$$

which is valid for spin-s fermions.

In a similar way we can obtain the Dyson equation for Green's function $G(X,Y)$:

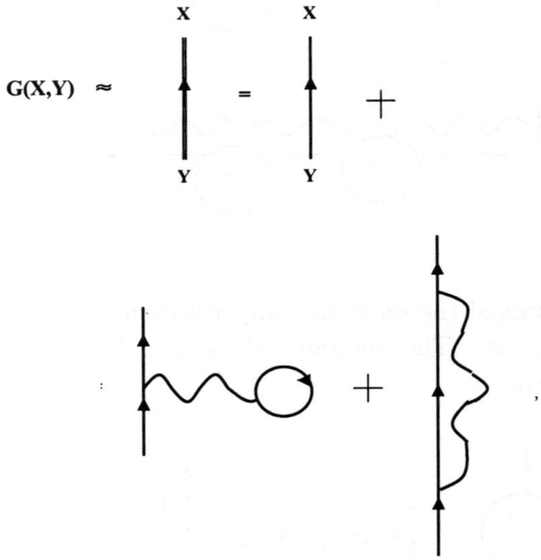

(11.6)

where as before, the light line denotes $G^{(0)}$, i.e. the noninteracting Green's function, and the bold line denotes the self-consistent Green's function G.

Chapter 12

The Correlation Energy of Electron Gas at High Density

Assume we are given an electron system at high enough density, that is the dimensionless parameter (without spin-interaction): $r_s = \dfrac{r_0}{a_0}$, where $\dfrac{4}{3}\pi r_0^3$ is the average volume occupied by a particle,

$$N\frac{4}{3}\pi r_0^3 = V, \qquad N/V = \rho\,,$$

and $a_0 = \dfrac{\hbar^2}{me^2}$ is the Bohr radius of a particle. If one defines the scaled quantities $\bar{V} \Longrightarrow r_0^{-3}V$, $\mathbf{k} \Longrightarrow r_0\mathbf{k}$, $\mathbf{p} \Longrightarrow r_0\mathbf{p}$, $\mathbf{q} \Longrightarrow \mathbf{q}r_0$, then one obtains the following dimensionless form of the Hamiltonian (5.1):

$$\hat{H} = \frac{e^2}{a_0 r_s^2}\left(\sum_{\mathbf{k},\alpha}\frac{|\mathbf{k}|^2}{2}a_{\mathbf{k}}^+ a_{\mathbf{k}} + \frac{r_s}{2\bar{V}}\sum_{\alpha,\beta,\mathbf{k},\mathbf{p},\mathbf{q}} U^{(2)}(\mathbf{q})a_{\mathbf{k}+\mathbf{q}_1,\alpha}^+ a_{\mathbf{p}-\mathbf{q},\beta}^+ a_{\mathbf{p},\beta}a_{\mathbf{k},\alpha}\right), \tag{12.1}$$

meaning that the perturbation term in (12.1) is tending to zero as $r_s \to 0$, corresponding to the high density limit $r_0 \to 0$. Unfortunately the dependence of the total energy E of the system (12.1) is not analytical in $r_s \in \mathbb{C}$:

$$E = \frac{Ne^2}{a_0 r_s^2}(a + br_s + cr_s^2 \ln_0 r_s + dr_s^2 + \cdots)\,, \tag{12.2}$$

where a, b, c and $d \in \mathbb{R}$ are *numerical* constants. To calculate first terms of (12.2), it is obviously enough to calculate Green's function $G_{\alpha\beta}(p)$, for the Hamiltonian (12.1), summing up the most divergent subseries corresponding to Feynman diagrams for the proper self-energy insertion Σ^*. These diagrams are called ladder ones and can be plotted as follows:

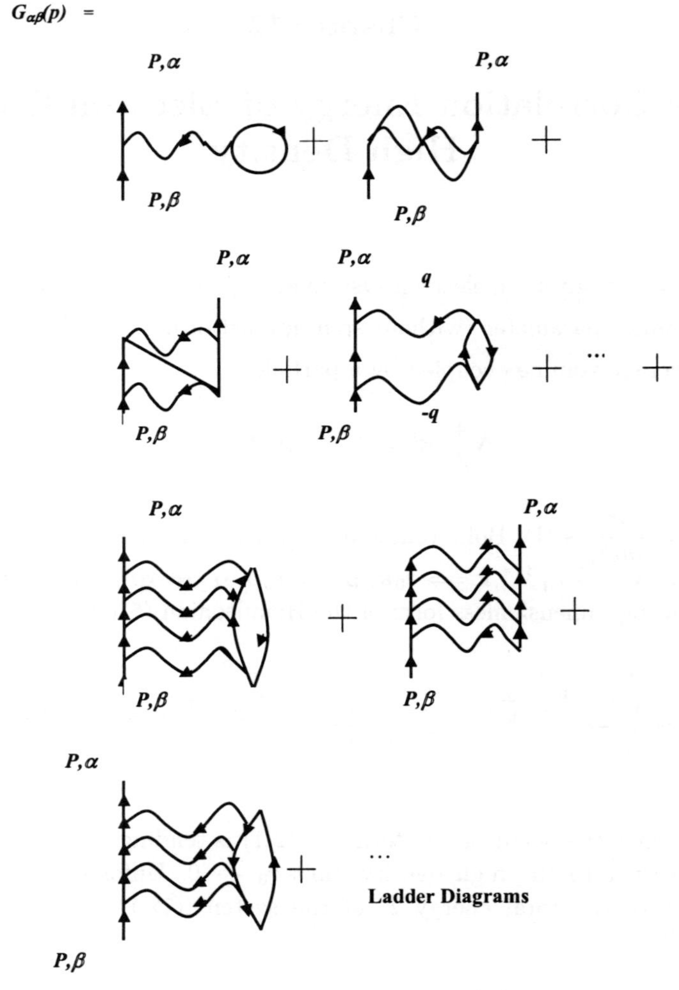

(12.3)

The corresponding proper self-energy Σ^* due to (12.3) can be represented as the following ladder approximation:

$$\hbar\Sigma^*(p) \simeq -2i \int \frac{d^4k}{(2\pi)^4} G^{(0)}(k)\Gamma(p,k;p,k) + i \int \frac{d^4k}{(2\pi)^4} G^{(0)}(k)\Gamma(k,p;p,k)\,, \tag{12.4}$$

where $\Gamma(p_1p_2;p_3p_4)$ is *an effective* two-particle interaction form, defined by the equation:

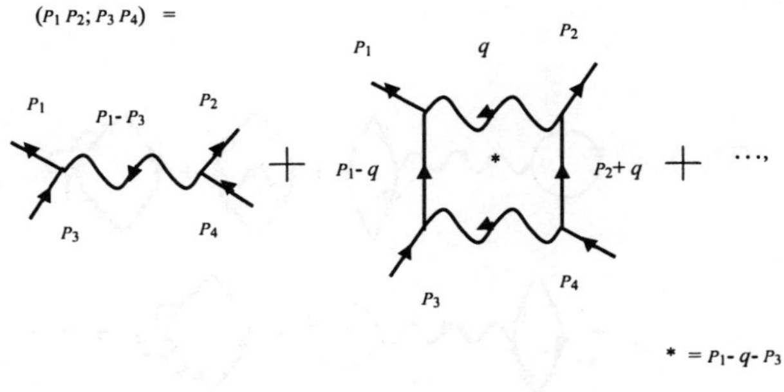

(12.5)

or diagrammatically as

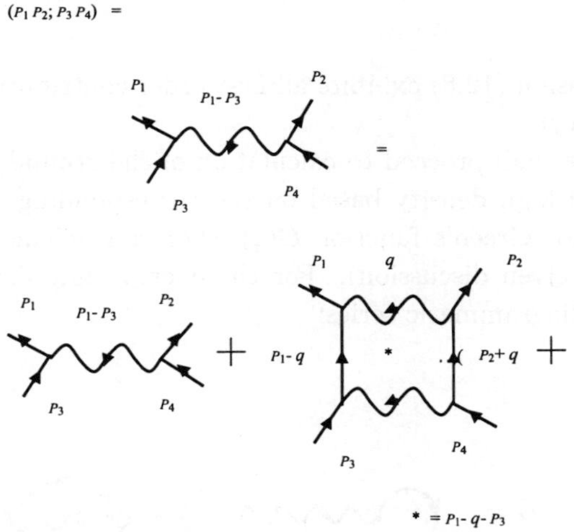

(12.6)

which is known as *a Bethe-Salpeter equation.* Similarly one can draw a graphical expression for the proper self-energy insertion $\Sigma^*(p)$ in ladder approximation:

$\hbar\Sigma^*(p) \simeq$ 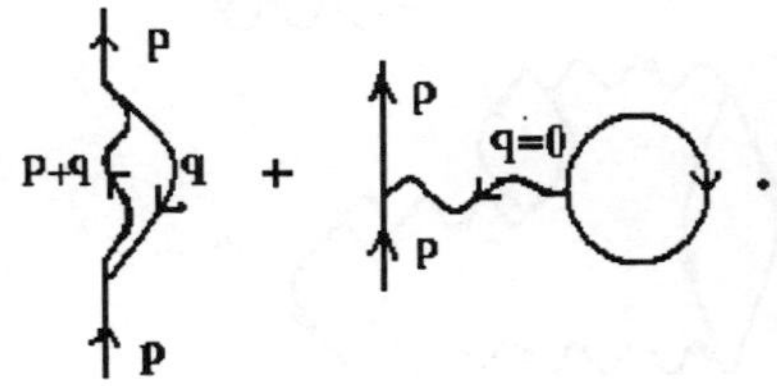

(12.7)

Subject to calculating the proper polarization insertion $\Pi^*(q)$ in (10.4) one can draw the following diagrammatic expression:

$\Pi^*(q) =$

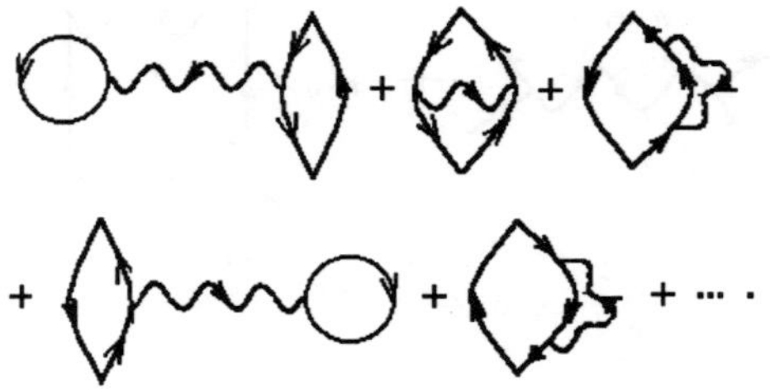

(12.8)

The expression (12.8) exhibits all first order contributions to proper polarization $\Pi^*(q)$.

Now we shall proceed to calculation of the ground state energy of electron gas of high density based on the corresponding diagrammatic representation for Green's function $G(q)$. (For convenience, $\hbar$ is suppressed in the below given discussion). For the energy shift $\triangle E$ one can draw the following diagrammatic series:

$\triangle E =$

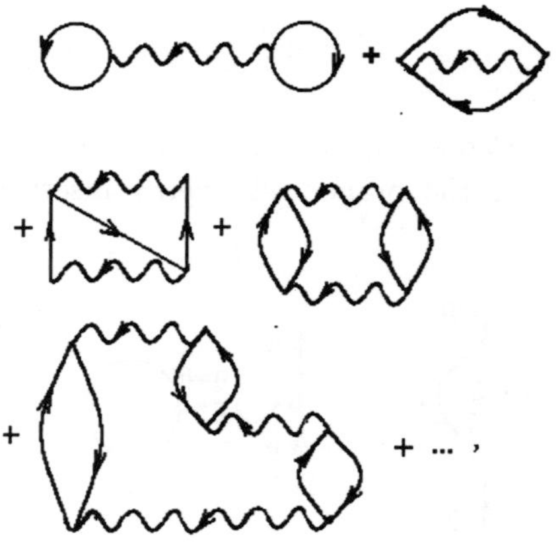

(12.9)

where in general, based on (5.20), one gets:

$$\triangle E = \sum_{n=1}^{\infty} \triangle E_{(n)} ,$$

$$\triangle E_{(n)} = \frac{(-1)^{(n+1+\bar{l}+\bar{h})}}{2^n} \times \sum_{\substack{\mathbf{ijkl},\ \mathbf{rsuv} \\ (connected \\ diagrams\ only!)}} \frac{\langle\ \mathbf{ij}|U^{(2)}|\mathbf{kl}\ \rangle \cdots \langle\ \mathbf{rs}|U^{(2)}|\mathbf{uv}\ \rangle}{(\triangle_2 + \triangle_3 + \cdots + \triangle_n)(\triangle_3 + \cdots \triangle_n) \cdots \triangle_n}, \tag{12.10}$$

with $\triangle_2 = \varepsilon_{\mathbf{i}} + \varepsilon_{\mathbf{j}} - \varepsilon_{\mathbf{k}} - \varepsilon_{\mathbf{l}}, \cdots \triangle_n = \varepsilon_{\mathbf{r}} + \varepsilon_{\mathbf{s}} - \varepsilon_{\mathbf{u}} - \varepsilon_{\mathbf{v}}$, for all $n \geq 1$, $\bar{l} \in \mathbb{Z}_+$ is the number of loops, $\bar{h} \in \mathbb{Z}_+$ is the number of hole lines.

The expression (12.10) can also be represented in the following form called the Reyleigh-Schrődinger form:

$$\triangle E = \sum_{n=1}^{\infty} \langle\ \Omega_0 | \hat{H}_{int} \left(\frac{1}{\mathbb{I}E_0 - \hat{H}_0} \hat{H}_{int} \right)^{n-1} |\Omega_0\ \rangle_{connected} , \tag{12.11}$$

being equivalent to the expression

$$\triangle E = \sum_{n=1}^{\infty} \langle\ \Omega_0 | \hat{H}_{int} \left[\frac{\hat{P}}{\mathbb{I}E_0 - \hat{H}_0} (E_0 - E - \hat{H}_{int})^{n-1} \right] |\Omega_0\ \rangle , \tag{12.12}$$

where the projection operator $\hat{P} : \Phi \longrightarrow \Phi$ along the state $|\Omega_0\ \rangle \in \Phi$ is given by the formula: $\hat{P} = \mathbb{I} - |\Omega_0\ \rangle\langle\ \Omega_0|$. ($\hat{P}|\Omega_0\ \rangle \equiv 0$!). The expression (12.12) is found from the following self-consistent equation on the ground state $|\Omega\rangle \in \Phi$

$$|\Omega\rangle = |\Omega_0\ \rangle + \frac{\hat{P}}{\mathbb{I}E_0 - \hat{H}_0} (E_0 - E - \hat{H}_{int}) |\Omega\rangle , \tag{12.13}$$

via the standard iteration.

Based on (12.9) we can now write down the following expression:

$$\triangle E_{(1)} = -\frac{4\pi e^2}{V} \sum_{\substack{\mathbf{p} \neq \mathbf{q} \\ |\mathbf{p}| \leq k_F \\ |\mathbf{q}| \leq k_F}} \frac{1}{|\mathbf{p} - \mathbf{q}|^2} \simeq \frac{2.21}{r_s^2} - \frac{0.916}{r_s} \tag{12.14}$$

-the so called exchange energy of Hartree-Fock type;

$$\triangle E_{(2)} = -\frac{32\pi^2 e^2 m}{V^2} \sum_{\substack{\mathbf{p},\mathbf{q},\mathbf{k} \\ |\mathbf{p}|,|\mathbf{q}|\leq k_F \\ |\mathbf{p}+\mathbf{k}|,|\mathbf{q}-\mathbf{k}| \rangle k_F}} \frac{1}{|\mathbf{k}|^4\langle\, \mathbf{k}, \mathbf{p}-\mathbf{q}+\mathbf{k}\,\rangle}$$

$$+\frac{16\pi^2 e^2 m}{V^2} \sum_{\substack{\mathbf{p},\mathbf{q},\mathbf{k} \\ |\mathbf{p}|,|\mathbf{q}|\leq k_F \\ |\mathbf{p}+\mathbf{k}|,|\mathbf{q}-\mathbf{k}| \rangle k_F}} \frac{1}{|\mathbf{k}|^2|\mathbf{p}-\mathbf{q}+\mathbf{k}|^2\langle\, \mathbf{k}, \mathbf{p}-\mathbf{q}+\mathbf{k}\,\rangle}$$

$$\simeq -\frac{3N}{8\pi^5}\int \frac{d^3\mathbf{k}}{|\mathbf{k}|^4}\int d^3\mathbf{q}\int \frac{d^3\mathbf{p}}{\langle\, \mathbf{k}, \mathbf{p}-\mathbf{q}+\mathbf{k}\,\rangle}$$

$$+\frac{3N}{16\pi^5}\int \frac{d^3\mathbf{k}}{|\mathbf{k}|^2}\int d^3\mathbf{q}\int \frac{d^3\mathbf{p}}{|\mathbf{p}-\mathbf{q}+\mathbf{k}|^2\langle\, \mathbf{k}, \mathbf{p}-\mathbf{q}+\mathbf{k}\,\rangle}$$

$$\simeq -\frac{3N}{8\pi^5}\int \frac{d^3\mathbf{k}}{|\mathbf{k}|^4}\int \frac{d^3\mathbf{q}}{\langle\, \mathbf{k}, \mathbf{p}-\mathbf{q}+\mathbf{k}\,\rangle} + 0.046N \tag{12.15}$$

Similarly one gets that

$$\triangle E^{ring}_{(3)} = \sum_{\substack{\mathbf{p}_1,\mathbf{p}_2,\mathbf{p}_3,\mathbf{k} \\ |\mathbf{p}_1|,|\mathbf{p}_2|,|\mathbf{p}_3|\leq k_F \\ |\mathbf{p}_1+\mathbf{k}|,|\mathbf{p}_2+\mathbf{k}|,|\mathbf{p}_3+\mathbf{k}| \rangle k_F}} \frac{8m^2(4\pi^2 e^2/V)^3}{|\mathbf{k}|^6\langle\, \mathbf{k}, \mathbf{p}_1+\mathbf{p}_2+\mathbf{k}\,\rangle\langle\, \mathbf{k}, \mathbf{p}_2+\mathbf{p}_3+\mathbf{k}\,\rangle}$$

$$= \frac{3N}{4\pi^5}\left(\frac{\alpha r_s}{\pi^2}\right)\int \frac{d^3\mathbf{k}}{|\mathbf{k}|^6}\int d^3\mathbf{p}_1\int \mathbf{d}^3\mathbf{p}_2\int$$

$$\times \frac{d^3\mathbf{p}_3}{\langle\, \mathbf{k}, \mathbf{p}_1+\mathbf{p}_2+\mathbf{k}\,\rangle\langle\, \mathbf{k}, \mathbf{p}_2+\mathbf{p}_3+\mathbf{k}\,\rangle}, \tag{12.16}$$

and other ones for arbitrary $n \in \mathbb{Z}_+$:

$$\triangle E^{ring}_{(n)} = (-1)^{n+1}\frac{3N}{8\pi^5}\left(\frac{\alpha r_s}{\pi^2}\right)^{n-2}\int I_n(|\mathbf{k}|)|\mathbf{k}|^{-2n}d^3\mathbf{k}, \tag{12.17}$$

where $\alpha = \left(\frac{4}{9\pi}\right)^{\frac{1}{3}}$, and

$$I_n(\mathbf{k}) = \frac{1}{n}\int_{-\infty}^{\infty} dt_1 \int_{-\infty}^{\infty} dt_2 \cdots \int_{-\infty}^{\infty} F_{\mathbf{k}}(t_1)F_{\mathbf{k}}(t_2)\cdots F_{\mathbf{k}}(t_n)$$

$$\times\, \delta(t_1+t_2+\cdots+t_n)dt_n\,,$$

$$F_{\mathbf{k}}(t) = \int d^3\mathbf{p} \exp\left[-|t|\left(\frac{1}{2}|\mathbf{k}|^2 + \langle\, \mathbf{k}, \mathbf{p}\,\rangle\right)\right] \tag{12.18}$$

Making use of the representation

$$\delta(x) = \frac{|\mathbf{k}|}{2\pi}\int \exp(i|\mathbf{k}|ux)du\,, \tag{12.19}$$

one can derive that

$$\begin{aligned}\triangle E^{ring}_{(n)} &= (-1)^{n+1}\frac{3N}{8\pi^5}\left(\frac{\alpha r_s}{\pi^2}\right)^{n-2} \\ &\times \frac{1}{2\pi n}\int \frac{d^3\mathbf{k}}{|\mathbf{k}|^{2n-1}}\int\limits_{-\infty}^{\infty}(4\pi R(u))^n du\,,\end{aligned} \tag{12.20}$$

$$R(u) = 1 - \arctan\frac{1}{u}\,.$$

Summing up all of the above contributions, one obtains the result for the correlation energy per electron:

$$\begin{aligned}&\varepsilon_{correlation} \\ &= 0.046 - \frac{3}{16\pi^6}\left(\frac{\pi^2}{\alpha r_s}\right)^2\int |\mathbf{k}|d^3\mathbf{k}\int\limits_{-\infty}^{\infty}\sum_{n\geq 2}^{\infty}\frac{(-1)^n}{n}\left[\frac{4R(u)\alpha r_s}{\pi|\mathbf{k}|^2}\right]^n du \\ &= 0.046 + \frac{3}{16\pi^6}\left(\frac{\pi^2}{\alpha r_s}\right)^2\int |\mathbf{k}|d^3\mathbf{k}\int\limits_{-\infty}^{\infty}\left[\ln\left\{1+\frac{4R(u)\alpha r_s}{\pi|\mathbf{k}|^2}\right\}-\frac{4R(u)\alpha r_s}{\pi|\mathbf{k}|^2}\right]du \\ &\simeq 0.046 + \frac{3}{16\pi^6}\left(\frac{\pi^2}{\alpha r_s}\right)^2\int\limits_{-\infty}^{\infty}du\int_0^1 4|\mathbf{k}|^3\pi\left[\ln\left(1+\frac{x}{|\mathbf{k}|^2}\right)-\frac{x}{|\mathbf{k}|^2}\right]d|\mathbf{k}|\,,\end{aligned} \tag{12.21}$$

where

$$x = \frac{4\alpha r_s R(u)}{\pi}\,. \tag{12.22}$$

Owing to the fact that $r_s \to 0$, (12.21) and (12.22) finalizes in:

$$\varepsilon_{correlation} \simeq \frac{2}{\pi^2}(1-\ln 2)\ln r_s + C_0 + O(r_s)\,, \tag{12.23}$$

where $C_0 \simeq -0.096$. Experiments give $C_0 \simeq -0.12$. The results above were produced by Gell-Mann and Brueckner in 1957. (Phys.Rev. 106 p. 364 (1957)).

PART 2

NONLINEAR QUANTUM OPTICS MODELS AND THEIR APPLICATIONS

Chapter 13

Nonlinear Quantum Optics Bose-Systems

We are now interested in a diagrammatic study of a Bose system described by the following simple Hamiltonian operator

$$\hat{H} = \sum_{s=\overline{1,2}} \hbar\omega_s a_s^+ a_s + i\chi(a_1^+ a_2^+ e^{-i\omega t} - a_1 a_2 e^{i\omega t}) \tag{13.1}$$

under the bose commutation conditions $[a_j, a_s^+] = \delta_{js}$ and $\omega = \omega_1 + \omega_2$. The Hamiltonian (13.1) models the process running in a down-conversion parametric amplifier of nonlinear optical devices.

The following fact is important for further analysis: the number operator $\hat{N} = \sum_{s=\overline{1,2}} a_s^+ a_s$ is not a conserved quantity of (13.1). Since the Hamiltonian (13.1) is time dependent, we can introduce the following simple change of variables:

$$a_s^+ :\longrightarrow a_s^+ e^{i\omega_s t}, \quad a_s :\longrightarrow a_s e^{-i\omega_s t} \tag{13.2}$$

to result in the expression

$$\hat{H} = \sum_{s=\overline{1,2}} \hbar\omega_s a_s^+ a_s + i\chi(a_1^+ a_2^+ - a_1 a_2)\,, \tag{13.3}$$

and take this as a starting point of our analysis.

Consider now the noninteracting Hamiltonian in (13.3), namely

$$\hat{H}_0 = \sum_{s=\overline{1,2}} \hbar\omega_s a_s^+ a_s\,, \tag{13.4}$$

possessing the following noninteracting photon ground state:

$$|\Omega_0\,\rangle = (a_2^+)^{N_2}|0\,\rangle\,, \tag{13.5}$$

consisting of N_2 "idler" photons with the energy $\hbar\omega_2$.

This state enjoys the following condition:

$$a_2^+ a_2 |\Omega_0\rangle = N_2 |\Omega_0\rangle , \tag{13.6}$$

since $[\hat{H}_0, a_s^+ a_s] = 0$, $s = \overline{1,2}$.

Now, taking into account the interaction $\hat{H}_{int} = \hat{H} - \hat{H}_0$, we are interested in the new ground state $|\Omega\rangle \in \Phi$ and in the law governing the average number of photons N_s, $s = \overline{1,2}$, in this state,

$$N_s = \langle \Omega | a_s^+ a_s | \Omega \rangle . \tag{13.7}$$

The quantity (13.7) can be calculated by means of Green's function as

$$N_s = G_{ss}(t; t^+) , \tag{13.8}$$

where, by definition

$$G_{sk}(t; t^+) = \frac{\langle \Omega | T(a_{s,H}(t) a_{k,H}^+(t^+)) | \Omega \rangle}{\langle \Omega | \Omega \rangle} , \tag{13.9}$$

for $k, s = \overline{1,2}$. (13.8) can be calculated in two ways: the first is due to the well known Bogoliubov Bose-condensation assumption, and the second one is due to Bogoliubov canonical transformation as the Hamiltonian (13.3) is in quadratic form. Let us illustrate these two ways:

1. Based on the definitions (13.5) and (13.6), and due to Bogoliubov's prescription, one can assume that

$$a_2 = \sqrt{N_2}\,\xi_2 \tag{13.10}$$

where, operators ξ_2 and $\xi_2^+ : \Phi \longrightarrow \Phi$ enjoy the relationship

$$[\xi_2, \ \xi_2^+] = \frac{1}{N_2}[a_2, a_2^+] = \frac{1}{N_2} . \tag{13.11}$$

Having supposed that $N_2 \to \infty$, that is the amount of photons in the "idler" subsystem characterized by the state $|\Omega_0\rangle \in \Phi$ is great enough, one can recognize that operators ξ_2, ξ_2^+ are equivalent to usual real numbers, as $[\xi_2, \ \xi_2^+] \simeq 0$ as $N_2 \to \infty$.

In the case of the interacting system (13.3), we need to impose the following external constraints on a possible ground state:

$$N_2 + \langle \Omega | \hat{N}_1 | \Omega \rangle = \bar{N} \tag{13.12}$$

where $\bar{N} \in \mathbb{Z}_+$ is the total fixed amount of all active nonlinear elements inside a crystal interacting with a laser radiation of circular frequency 2ω.

Introducing a so called Lagrangian multiplier into the Hamiltonian (13.3), the constraint (13.12) gets involved with our analysis self-consistently:

$$\hat{H} \to \hat{H}_\mu = \sum_{s=\overline{1,2}} \hbar(\omega_s - \mu) a_s^+ a_s + i\chi(a_1^+ a_2^+ - a_1 a_2)\,, \tag{13.13}$$

where $\mu \in \mathbb{R}$ is known as a chemical potential.

Thus, our problem is to find the ground state vector $|\Omega_\mu\,\rangle \in \Phi$, depending on $\mu \in \mathbb{R}$, and realizing the consistency condition with (13.12). To make this a more accurate, let us consider the equation for the ground state $|\Omega_\mu\,\rangle$ of the Hamiltonian (13.13):

$$\hat{H}_\mu |\Omega_\mu\,\rangle = E_\mu |\Omega_\mu\,\rangle\,, \tag{13.14}$$

satisfying the condition

$$N_2 + \langle\, \Omega_\mu | \hat{N}_1 | \Omega_\mu\,\rangle = \bar{N} \tag{13.15}$$

stemming from (13.12). From the definition (13.14) and (13.15) one derives the following criteria right away:

$$-\frac{\partial E_\mu}{\partial \mu} = \bar{N} \tag{13.16}$$

holds at some value of $\mu \in \mathbb{R}$. Thereby, equation (13.16) will serve for us as a determining one of the chemical potential $\mu \in \mathbb{R}$. Having found $\mu \in \mathbb{R}$, there is now a problem of concern with (13.16) and that is the determination of the amount of photons generated by this optical device due to the interaction of a crystal with a laser radiation:

$$N_1 = \langle\, \Omega_\mu | a_1^+ a_1 | \Omega_\mu\,\rangle \tag{13.17}$$

at $\mu \in \mathbb{R}^1$ calculated from (13.16).

To proceed, let us put $\xi_2 = \xi_2^+ = 1$ in (13.13). This gives rise to

$$\hat{H}_\mu \simeq (\hbar\omega_1 - \mu) a_1^+ a_1 - (\hbar\omega_2 - \mu) N_2 + i\chi\sqrt{N_2}(a_1^+ - a_1)\,. \tag{13.18}$$

The quadratic form (13.18) can easily be diagonalized as:

$$\tilde{H}_\mu \simeq (\hbar\omega_1 - \mu)\tilde{a}_1^+ \tilde{a}_1 + N_2(\hbar\omega_2 - \mu)\,, \tag{13.19}$$

where, by definition we put

$$a_1 = \tilde{a}_1 - \frac{i\chi\sqrt{N_2}}{\omega_1\hbar - \mu}, \quad a_1^+ = \tilde{a}_1^+ + \frac{i\chi\sqrt{N_2}}{\omega_1\hbar - \mu}\,. \tag{13.20}$$

Now there is no problem to construct the eigenstate $|\tilde{\Omega}_\mu\rangle \in \Phi$ for the Hamiltonian (13.19):

$$|\tilde{\Omega}_\mu\rangle = \frac{1}{(\bar{N}_1!)^{1/2}}(\tilde{a}^+)^{\bar{N}_1}|\tilde{0}_\mu\rangle, \tag{13.21}$$

where $|\tilde{0}_\mu\rangle \in \Phi$ is the corresponding vacuum state, and $\bar{N}_1 \in \mathbb{Z}_+$ is some integer still not specified. The eigenvalue $E_\mu \in \mathbb{R}^1$, satisfying (13.14) is now obtained easily as

$$E_\mu = (\omega_1\hbar - \mu)\bar{N}_1 + (\omega_2\hbar - \mu)N_2 + N_2\frac{[(\omega_1\hbar - \mu)^2 - \chi^2]}{(\hbar\omega_1 - \mu)}. \tag{13.22}$$

We still need to impose the constraint (13.15) which can be found based on transformation formulas (13.20) as

$$N_2 + \langle\tilde{\Omega}_\mu|\left(\tilde{a}_1^+ + \frac{i\chi\sqrt{N_2}}{\omega_1\hbar - \mu}\right)\left(\tilde{a}_1 - \frac{i\chi\sqrt{N_2}}{\omega_1\hbar - \mu}\right)|\tilde{\Omega}_\mu\rangle$$

$$= \frac{N_2 + \bar{N}_1 + N_2\chi^2}{(\hbar\omega_1 - \mu)^2} = \bar{N} \tag{13.23}$$

From (13.22) and (13.23) we get the expression

$$E_\mu = (\omega_1\hbar - \mu)\left(\frac{\bar{N} - N_2 - N_2\chi^2}{(\hbar\omega_1 - \mu)^2}\right) + (\omega_2\hbar - \mu)N_2 + N_2\frac{[(\omega_1\hbar - \mu)^2 - \chi^2]}{(\omega_1\hbar - \mu)}$$

$$= \bar{N}(\hbar\omega_1 - \mu) + \frac{N_2}{\hbar\omega_1 - \mu}\left[(\omega_1\hbar - \mu)(\omega_2\hbar - \mu) - 2\chi^2\right]. \tag{13.24}$$

Having imposed upon (13.24) the condition

$$\frac{\partial E_\mu}{\partial \mu} = \bar{N}, \tag{13.25}$$

one obtains

$$\frac{(\omega_1\hbar - \mu)(\omega_2\hbar - \mu) - 2\chi^2}{(\omega_1\hbar - \mu)} = k \tag{13.26}$$

as a constant parameter. To determine k explicitly, let us make use of the expression (13.23)

$$\bar{N}_1 + \frac{N_2[(\omega_1\hbar - \mu)^2 + \chi^2]}{(\omega_1\hbar - \mu)^2}$$

$$= \bar{N}_1 + \frac{N_2}{(\omega_1\hbar - \mu)^2}\left\{\hbar(\omega_1 - \mu)^2 + [(\omega_1\hbar - \mu)(\omega_2\hbar - \mu) - k(\omega_1\hbar - \mu)]\frac{1}{2}\right\}$$

$$= \bar{N}_1 + \frac{N_2}{1}\left\{1 + \frac{\omega_2\hbar - \mu}{2(\omega_1\hbar - \mu)} - \frac{k}{2(\omega_1\hbar - \mu)}\right\}$$

$$= \bar{N}_1 + \frac{\{2\omega_1\hbar - 2\mu + \omega_2\hbar - \mu - k\}N_2}{2(\omega_1\hbar - \mu)}$$

$$= \bar{N}_1 + \frac{\{2\omega_1\hbar + \omega_2\hbar - 3\mu - k\}N_2}{2(\omega_1\hbar - \mu)} \Longrightarrow \bar{N}\,. \tag{13.27}$$

Further, let us note here that the parameter $k \in \mathbb{R}_+$, depends in general on ω_1, ω_2, χ and $\bar{N}/N_2$. This dependence can easily be retrieved from the condition that the energy E_μ (13.25) achieves its minimum subject to $N_2 \in \mathbb{Z}_+$:

$$\frac{\partial E_\mu(N_2, \mu(N_2))}{\partial N_2} = 0 \tag{13.28}$$

Making use of (13.24) and (13.26), the condition (13.25) gives the result:

$$k = \frac{c(\chi; \omega_1, \omega_2)\bar{N}}{N_2} \tag{13.29}$$

where $c(\chi; \omega_1, \omega_2)$ is some parameter depending on crystal data χ, ω_1 and ω_2. Another important point is to be mentioned now concerning the integer parameter $\bar{N}_1 \in \mathbb{Z}_+$ entering the formula (13.27). As the parameter $\bar{N}_1 \in \mathbb{Z}_+$ is integer, it can not depend on the coupling interaction parameter χ continuously. This obviously means that $\bar{N}_1 \in \mathbb{Z}_+$ can be chosen in some sense arbitrarily depending only on the parameter $c(\chi; \omega_1, \omega_2)$ and some subsidiary physical conditions involved.

For instance, let us consider the case when $\bar{N}_1 = 0$. This clearly means that the crystal possesses no ω_1–photon inside the crystal before interaction. As a result, from (13.27) one gets

$$N_2 = \frac{\bar{N}\ 2(\hbar\omega_1 - \mu)}{(2\omega_1\hbar + \omega_2\hbar - 3\mu - k)}\,, \tag{13.30}$$

or, based on (13.23),

$$N_1 = \bar{N} - N_2 = \frac{N_2\chi^2}{(\hbar\omega_1 - \mu)^2}$$

$$= \frac{\bar{N}\ 2\chi^2}{(2\omega_1\hbar + \omega_2\hbar - 3\mu - k)(\hbar\omega_1 - \mu)} \simeq \frac{\bar{N}\chi^2}{(\hbar\omega_1 - \hbar\omega_2)^2} \tag{13.31}$$

where we took into account that $\mu \simeq \omega_2$ has $\chi \simeq 0$. The latter means that $c \to 0$ as $\chi \to 0$, or more accurately.

$$c \simeq \frac{2\chi^2}{\hbar(\omega_2 - \omega_1)} \tag{13.32}$$

Then (13.29) becomes

$$k \simeq \frac{2\chi^2 \bar{N}}{\hbar(\omega_2 - \omega_1)N_2} \,. \tag{13.33}$$

Substituting (13.33) into (13.26)

$$\frac{(\omega_1\hbar - \mu)(\omega_2\hbar - \mu) - 2\chi^2}{\omega_1\hbar - \mu} \simeq \frac{2\chi^2}{\hbar(\omega_2 - \omega_1)}. \tag{13.34}$$

This expression can be solved approximately by taking $\hbar^{-1}\mu = \omega_2 + \chi\omega_2^{(1)} + \chi^2\omega_2^{(2)} + O(\chi^3)$. As a result, $\omega_2^{(1)} = 0$, $\omega_2^{(1)} = 0$, or $\mu \simeq \omega_2$ up to $O(\chi^3)$ as $\chi \to 0$. So, the interaction of the crystal active centers with an external laser radiation gives rise to generation of N_1 quanta of photons with frequency ω_1, where owing to (13.31),

$$N_1 \simeq \frac{\bar{N}\chi^2}{(\omega_2 - \omega_1)^2\hbar^2} \tag{13.35}$$

These quanta fill in a signal cavity inside the crystal, possessing at the same time an idler bath of quanta with frequency ω_1 and serving as a stimulator of the generation of ω_1–quanta inside the crystal.

2. We shall next discuss the method of calculating the signal quanta (13.35) based on Green's function approach and the associated Feynman diagram representation of interaction processes taking place inside a crystal and described by the Hamiltonian (13.15) with the constraint (modifying (13.12))

$$\langle\, \Omega_\mu | a_1^+ a_1 | \Omega_\mu \,\rangle + \langle\, \Omega_\mu | a_2^+ a_2 | \Omega_\mu \,\rangle = \bar{N} \tag{13.36}$$

where the number $\bar{N} \in \mathbb{Z}_+$ is fixed as before.

To proceed and to be more accurate let us rewrite the Hamiltonian (13.1) in the following complete form:

$$\hat{H} = \hbar\omega_1 a_1^+ a_1 + \hbar\omega_2 a_2^+ a_2 + 2\hbar\omega_0 c^+ c_1 + i\chi(a_1^+ a_2^+ c - a_1 a_2 c^+)\,, \tag{13.37}$$

where c and c^+ are destruction and creation operators of the external laser radiation, exerting the influence on photon states inside the crystal. At $c \simeq \exp(i2\omega_0 t)$ (13.37) leads obviously to (13.1).

For the Feynman perturbation technique to be activated it is necessary to build the noninteracting ground state $|\Omega_{0,\mu}\ \rangle \in \Phi$. The perturbation Hamiltonian $\hat{H}_{int}$ in the interaction representation will be constructed subject to this ground state with effective creation and destruction operators, all satisfying the fundamental condition

$$\tilde{a}_1|\Omega_{0,\mu}\ \rangle = 0, \quad \tilde{a}_2|\Omega_{0,\mu}\ \rangle = 0, \quad \tilde{c}|\Omega_{0,\mu}\ \rangle = 0\,. \tag{13.38}$$

Let us now introduce this procedure: we define the operators

$$a_2 = \sqrt{N_2} + \tilde{a}_2, \quad a_1 = \tilde{a}_1, \quad c = \tilde{c},\ [\tilde{a}_1, \tilde{a}_1^+] = 1 = [\tilde{a}_2, \tilde{a}_2^+]\,, \tag{13.39}$$

and substitute them into (13.37) to obtain

$$\begin{aligned} \hat{H} &= \hbar\omega_1 \tilde{a}_1^+ \tilde{a}_1 + \hbar\omega_2 \tilde{a}_2^+ \tilde{a}_2 + 2\hbar\omega_2 N_2 + \hbar\omega_2 \sqrt{N_2}(\tilde{a}_2^+ + \tilde{a}_2) \\ &\quad + i\chi[(\tilde{a}_1^+ \tilde{a}_2^+ \tilde{c} - \tilde{a}_1 \tilde{a}_2 \tilde{c}^+) + \sqrt{N_2}(\tilde{a}_1^+ \tilde{c} - \tilde{a}_1 \tilde{c}^+)] + 2\hbar\omega_0 \tilde{c}^+ \tilde{c}^+\,. \end{aligned} \tag{13.40}$$

The condition (13.36) is taken into account by introducing chemical potentials $\mu_0,\ \mu \in \mathbb{R}^1$ into (13.40) $\hat{H} \to \hat{H}_\mu$, where

$$\begin{aligned} \hat{H}_\mu &= \tilde{a}_1^+ \tilde{a}_1(\hbar\omega_1 - \mu) + \tilde{a}_2^+ \tilde{a}_2(\hbar\omega_2 - \mu) + N_2(\hbar\omega_2 - \mu) \\ &\quad + \sqrt{N_2}(\tilde{a}_2^+ + \tilde{a}_2)(\hbar\omega_2 - \mu) + i\chi[(\tilde{a}_1^+ \tilde{a}_2^+ \tilde{c} - \tilde{a}_1 \tilde{a}_2 \tilde{c}^+) \\ &\quad + \sqrt{N_2}(\tilde{a}_1^+ \tilde{c} - \tilde{a}_1 \tilde{c}^+)] + (2\hbar\omega_0 - \mu_0)\tilde{c}^+ \tilde{c}^+\,, \end{aligned} \tag{13.41}$$

under the extended condition (13.36)

$$\langle\ \Omega_\mu|\tilde{a}_1^+ \tilde{a}_1|\Omega_\mu\ \rangle + N_2 = \bar{N},\ \langle\ \Omega_\mu|\tilde{c}^+ \tilde{c}|\Omega_\mu\ \rangle = \bar{N}_0\,. \tag{13.42}$$

The ground state $|\Omega_\mu\ \rangle \in \Phi$ satisfies the following Schrödinger equation:

$$\hat{H}_\mu|\Omega_\mu\ \rangle = E_\mu|\Omega_\mu\ \rangle\,, \tag{13.43}$$

with the eigenvalue $E_\mu \in \mathbb{R}^1$ determining chemical potentials $\mu_0,\ \mu \in \mathbb{R}^1$:

$$-\frac{\partial E_\mu}{\partial \mu} = \bar{N},\ -\frac{\partial E_\mu}{\partial \mu_0} = \bar{N}_0\,. \tag{13.44}$$

Subject to the noninteracting ground state $|\Omega_{0,\mu}\ \rangle \in \Phi$ we suppose that the following standard conditions are enjoyed:

$$\begin{aligned} &\tilde{a}_1|\Omega_{0,\mu}\ \rangle = 0,\ \tilde{a}_2|\Omega_{0,\mu}\ \rangle = 0, \quad \tilde{c}|\Omega_{0,\mu}\ \rangle = 0\,, \\ &\hat{H}_{0,\mu}|\Omega_{0,\mu}\ \rangle = E_{0,\mu}|\Omega_{0,\mu}\ \rangle\,, \end{aligned} \tag{13.45}$$

where, due to (13.42)

$$E_{0,\mu} = \langle\, \Omega_{0,\mu} | \hat{H}_{0,\mu} | \Omega_{0,\mu}\, \rangle = N_2(\hbar\omega_2 - \mu)\,. \tag{13.46}$$

The above assumptions obviously mean that the noninteracting ground state $|\Omega_{0,\mu}\,\rangle \in \Phi$ is equivalent to a vacuum state for the Hamiltonian (13.41) shifted by a constant term. Thereby we arrive at a position suitable for application of the Feynman diagrammatic tools.

As we know, the main ingredient of Feynman's perturbation technique is the Green's functions $G^{(0)}(t,t')$ and $G^{(0)}_{ij}(t,t')$ of the noninteracting Hamiltonian $\hat{H}_{0,\mu} = \hat{H}_\mu|_{\chi=0}$, defined as

$$\begin{aligned} iG^{(0)}_{ij}(t,t') &= \langle\, \Omega_{0,\mu} | T(\tilde{a}_{i,I}(t)\tilde{a}^+_{j,I}(t')) | \Omega_{0,\mu}\, \rangle\,, \\ iG^{(0)}(t,t') &= \langle\, \Omega_{0,\mu} | T(\tilde{c}(t)\tilde{c}^+(t')) | \Omega_{0,\mu}\, \rangle\,, \end{aligned} \tag{13.47}$$

where $i,j = \overline{1,2}$, and

$$\begin{aligned} \tilde{c}(t) &= \exp\left(\frac{i}{\hbar}\hat{H}_{0,\mu}t\right)\tilde{c}\exp\left(-\frac{i}{\hbar}\hat{H}_{0,\mu}t\right), \\ \tilde{a}_i(t) &= \exp\left(\frac{i}{\hbar}\hat{H}_{0,\mu}t\right)\tilde{a}_i\exp\left(-\frac{i}{\hbar}\hat{H}_{0,\mu}t\right), \\ \tilde{a}^+_i(t) &= \exp\left(\frac{i}{\hbar}\hat{H}_{0,\mu}t\right)\tilde{a}^+_i\exp\left(-\frac{i}{\hbar}\hat{H}_{0,\mu}t\right). \end{aligned} \tag{13.48}$$

Based on the definitions (13.45), one finds easily that

$$\begin{aligned} \tilde{a}_1(t) &= \exp\left(-ti\left(\omega_1 - \frac{\mu}{\hbar}\right)\right)\tilde{a}_1\,, \\ \tilde{a}_2(t) &= \exp\left(-i\left(\omega_2 - \frac{\mu}{\hbar}\right)t\right)\left[\tilde{a}_2 - \sqrt{N_2}\left(e^{i\left(\omega_2 - \frac{\mu}{\hbar}\right)t} - 1\right)\right], \\ \tilde{c}(t) &= \exp\left(-i\left(2\omega_0 - \frac{\mu}{\hbar}\right)t\right)\tilde{c}\,, \end{aligned} \tag{13.49}$$

giving rise (owing to (13.47)) to the following exact Green's functions:

$$G^{(0)}_{11}(t,t') = \int\limits_{\mathbb{R}} \frac{d\omega\exp(-i\omega(t-t'))}{2\pi\left(\omega - \omega_1 + \frac{\mu}{\hbar} + i\eta\right)}\,; \tag{13.50}$$

$$G^{(0)}_{12}(t,t') = 0, \quad G^{(0)}_{21}(t,t') = 0\,; \tag{13.51}$$

$$G_{22}^{(0)}(t,t') = \frac{1}{2\pi}\int\limits_{\mathbb{R}} \frac{d\omega \exp(-i\omega(t-t'))}{\left(\omega - \omega_2 + \frac{\mu}{\hbar} + i\eta\right)} + N_2 e^{-i(t-t')(\omega_2 - \mu/\hbar)}$$
$$\times (e^{it(\omega_2-\mu/\hbar)} - 1)(e^{it'(\omega_2-\mu/\hbar)} - 1)\,; \qquad (13.52)$$

$$G_0^{(0)}(t,t') = \frac{1}{2\pi}\int\limits_{\mathbb{R}} \frac{d\omega \exp(-i\omega(t-t'))}{\left(\omega - 2\omega_0 + \frac{\mu}{\hbar} + i\eta\right)}\,. \qquad (13.53)$$

We can now construct proper self-energy kernels $\Sigma_{i,j}^*$, Σ_j^* and Σ_0^* defining the full Green's functions

$$\begin{aligned} iG_{ij}^{(0)}(t,t') &= \langle\, \Omega_\mu | T(\tilde{a}_{i,H}(t)\tilde{a}_{j,H}^{+}(t'))|\Omega_\mu\,\rangle\,, \\ iG^{(0)}(t,t') &= \langle\, \Omega_\mu | T(\tilde{c}_{H}(t)\tilde{c}_{H}^{+}(t'))|\Omega_\mu\,\rangle\,, \qquad (13.54) \\ iG_{j}^{(0)}(t,t') &= \langle\, \Omega_\mu | T(\tilde{a}_{j,H}(t)\tilde{c}_{H}^{+}(t'))|\Omega_\mu\,\rangle \end{aligned}$$

making use of the Dyson expressions

$$\begin{aligned} G_{ij}(t,t') &= G_{ij}^{(0)}(t,t') + \int d\tau \int d\tau' G_{ik}^{(0)}(t,\tau) \sum\nolimits_{ks}^{*}(\tau,\tau')G_{sj}(\tau,t')\,, \\ G_{0}(t,t') &= G_{0}^{(0)}(t,t') + \int d\tau \int d\tau' G_{0}^{(0)}(t,\tau) \sum\nolimits_{0}^{*}(\tau,\tau')G_{0}(\tau,t')\,, \qquad (13.55) \\ G_{j}(t,t') &= G_{j}^{(0)}(t,t') + \int d\tau \int d\tau' G_{j}^{(0)}(t,\tau) \sum\nolimits_{j}^{*}(\tau,\tau')G_{j}(\tau,t')\,. \end{aligned}$$

Namely, we get the following diagrammatic expansions for proper self-energies:

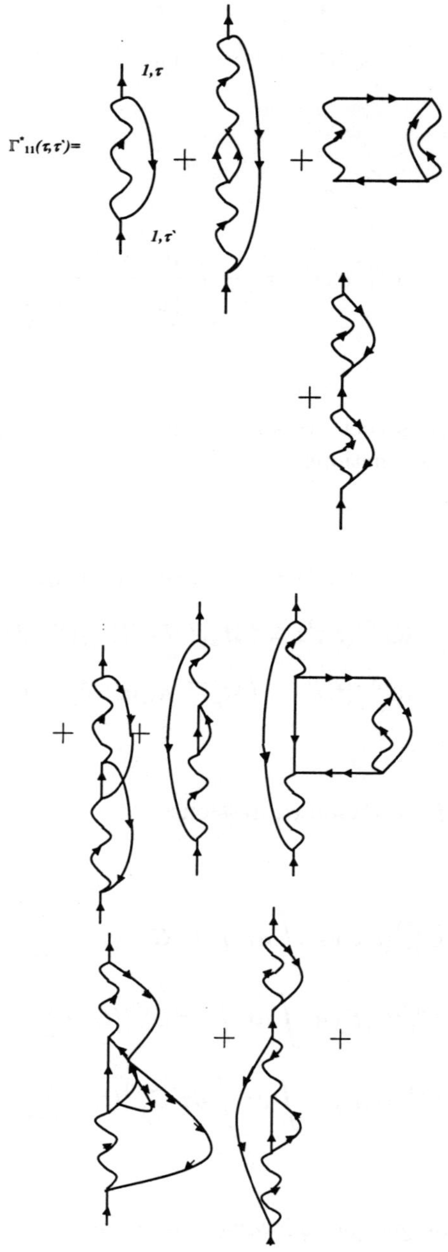

(13.56)

$\Sigma_{22}^{*}(\tau, \tau') =$

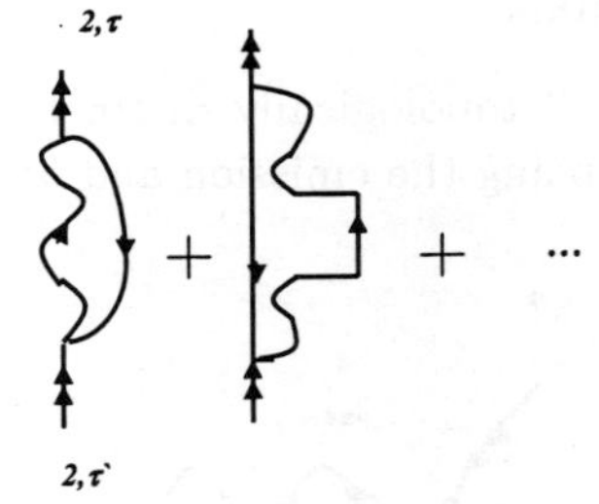

$\Sigma_{12}^{*}(\tau, \tau') =$

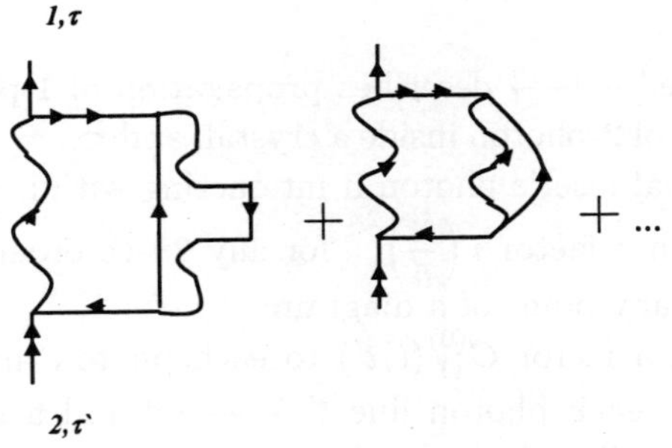

$\Sigma_{21}^{*}(\tau, \tau') =$

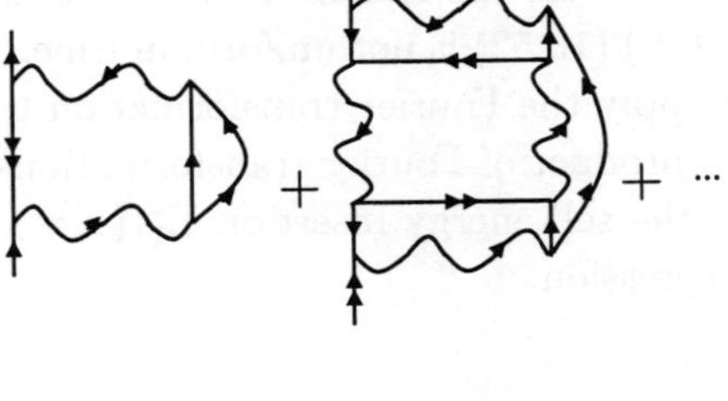

and a similar diagrammatic series for Σ_j^*, $j = \overline{1,2}$. Based on general prescriptions subject to Feynman's diagrams (13.56), the following rules should be used when calculating self-energies.

Feynman'sRules :

1. Construct all topologically distinct connected Feynman diagrams, the basic vertices being the emission and absorptions of photons:

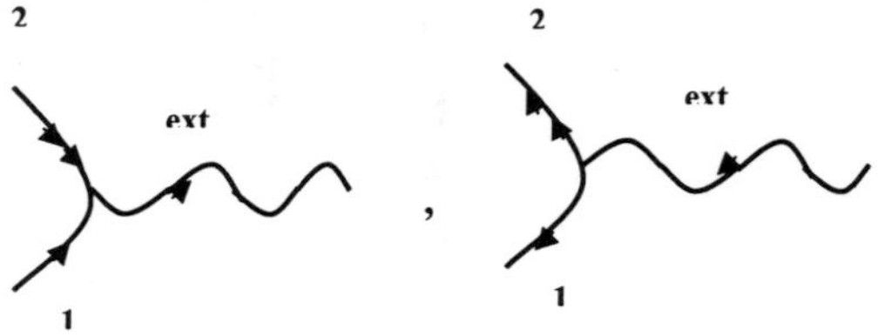

(13.57)

where a line $\longrightarrow$ describes propagation of 1-photon, a line $\longrightarrow\!\!\!\!\!\!\!\!\longrightarrow$ describes that of 2-photon inside a crystal, and $\sim\!\!\!\rightarrow\!\!\!\sim$ describes propagation of an external laser's photon if interacting with a crystal;

2. Assign a factor $i\left(\frac{\chi}{\hbar}\right)^{2n}$ for any $2n$-th order in perturbation expansion, or for any point of a diagram;

3. Affix a factor $G^{(0)}_{11}(t,t')$ to each photon line $t' \longrightarrow t$, a factor $G^{(0)}_{22}(t,t')$ to each photon line $t' \longrightarrow\!\!\!\!\!\!\!\!\longrightarrow t$ and a factor $G^{(0)}_{0}(t,t')$ to each laser's photon line $t' \sim\!\!\!\rightarrow\!\!\!\sim t$;

4. Integrate over all internal times t_i, $i = \overline{1, 2n-2}$.

As an example, let us now consider the Dyson equation for the Green's function $G^{(0)}_{11}(t,t')$:

$$G_{11}(t,t') = G^{(0)}_{11}(t,t')G^{(0)}_{11}(t,\tau) * \Sigma^{*}_{11}(\tau,\tau') * G_{11}(\tau',t')\,, \tag{13.58}$$

where the sign "$*$" is for the convolution operation in (13.55). Since the momentum operator $\hat{P}$ for the Hamiltonian (13.40) is not defined, the Green's function $G^{(0)}_{22}(t,t')$ (13.52) is not uniform in time. The latter does not make it possible to apply the Fourier transformation to (13.58) and write down the result as a product of Fourier transformations.

Subject to the self-energy insertion $\Sigma^{*}_{11}(\tau,\tau')$ based on (13.56) one arrives at the expression:

$$\begin{aligned}
&\Sigma^{*}_{11}(\tau,\tau')\\
&= i\left(\frac{\chi}{\hbar}\right)^{2} G^{(0)}_{22}(\tau',\tau)G^{(0)}_{0}(\tau,\tau')\\
&+ i\left(\frac{\chi}{\hbar}\right)^{4} G^{(0)}_{22}(\tau',\tau)\int dt_1 \int dt_2 G^{(0)}_{0}(\tau,t_1)G^{(0)}_{11}(t,t_1)G^{(0)}_{22}(t,t_1)G^{(0)}(t_2,\tau')
\end{aligned}$$

$$+i\left(\frac{\chi}{\hbar}\right)^4 G_0^{(0)}(\tau',\tau)\int dt_1\int dt_2 G_{22}^{(0)}(t_1,\tau)G_{11}^{(0)}(t,t_1)G_0^{(0)}(t_2,t_1)G_{22}^{(0)}(t_2,\tau')$$

$$+i\left(\frac{\chi}{\hbar}\right)^6\int dt_1\int dt_2\int dt_3\int dt_4 G_{22}^{(0)}(\tau,t_1)G_{11}^{(0)}(t_1,t_2)G_{22}^{(0)}(t_3,t_2)$$

$$\times G_0^{(0)}(t_3,t_2)G_{11}^{(0)}(t_3,t_4)G_0^{(0)}(t_4,\tau')G_{22}^{(0)}(\tau',\tau)+\cdots \tag{13.59}$$

Having calculated all self-energy insertions (13.56) and then solved equations (13.55), one can easily enjoy the constraints (13.42) as follows:

$$\begin{aligned}\lim\nolimits_{\varepsilon\downarrow 0} G_{11}(t,t+\varepsilon)+N_2&=\bar{N}\\ \lim\nolimits_{\varepsilon\downarrow 0} G_0(t,t+\varepsilon)+N_2&=\bar{N}_0\,.\end{aligned} \tag{13.60}$$

Making use of equations (13.44), where the energy

$$E_\mu=(\hbar\omega_1-\mu)\lim_{\varepsilon\downarrow 0}G_{11}(t,t+\varepsilon)+(\hbar\omega_2-\mu)\lim_{\varepsilon\downarrow 0}G_{22}(t,t+\varepsilon)$$

$$+(2\hbar\omega_0-\mu_0)\lim_{\varepsilon\downarrow 0}G_0(t,t+\varepsilon)+i\chi\langle\,\Omega_\mu|(\tilde{a}_1^+\tilde{a}_2^+\tilde{c}-\tilde{a}_1\tilde{a}_2\tilde{c}^+)|\Omega_\mu\,\rangle$$

$$+N_2((\hbar\omega_2-\mu)+i\chi\sqrt{N_2}[G_1^*(t,t+\varepsilon)-G_1(t,t+\varepsilon)] \tag{13.61}$$

can be calculated using, for instance, the following diagrammatic expansion for terms $\langle\, \Omega_\mu | \tilde{a}_1^+ \tilde{a}_2^+ \tilde{c} | \Omega_\mu \,\rangle$ and $\langle\, \Omega_\mu | \tilde{a}_1 \tilde{a}_2 \tilde{c}^+ | \Omega_\mu \,\rangle$:

$$\langle\, \Omega_\mu | \tilde{a}_1^+ \tilde{a}_2^+ \tilde{c} | \Omega_\mu \,\rangle = \tag{13.62}$$

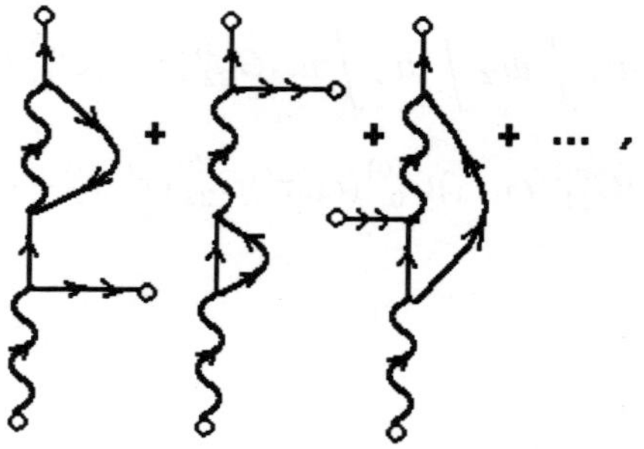

together with $\langle\, \Omega_\mu | \tilde{a}_1 \tilde{a}_2 \tilde{c}^+ | \Omega_\mu \,\rangle = (\langle\, \Omega_\mu | \tilde{a}_1^+ \tilde{a}_2^+ \tilde{c} | \Omega_\mu \,\rangle)^*$. Thus, the equations (13.60)-(13.62) constitute the full system for determining the chemical potentials $\mu_0,\ \mu \in \mathbb{R}^1$ subject to the constraints (13.60). Having taken into account that a parameter $\chi \in \mathbb{R}_+$ is too small, we can easily compute all Green's functions (13.54) and find not only the sought amount $\bar{N}_1 = \langle\, \Omega_\mu | \tilde{a}_1^+ \tilde{a}_1 | \Omega_\mu \,\rangle$ of photons in the signal due to the interaction of a crystal with a laser radiation considered in our model, but also convince ourselves that the result obtained will coincide almost completely with (13.31) found before via the Bogoliubov canonical transformation method.

Chapter 14

The Dicke Model

The Dicke model is one of the simplest models describing a coherent monochromatic (laser) electromagnetic radiation. To start with, consider a nonrelativistic electron in a radiation field (external) specified by an electromagnetic vector potential $\mathbf{A} \in \mathbb{R}^3$:

$$\hat{H}_{\mathbf{k}} = \frac{1}{2m} \, |\hbar \mathbf{k} - \frac{e}{c} \mathbf{A}|^2 + \frac{|\mathbf{E}|^2 + |\mathbf{B}|^2}{8\pi} \,,$$

where $\mathbf{k} \in \mathbb{R}^3$ is the electron momentum. Since $\frac{e^2}{2mc^2}$ is too small a quantity, in the first order of perturbation theory we may neglect the term $\frac{|\mathbf{A}|^2 e^2}{2mc^2}$. Let us consider that the electron belongs to an atom of *active medium* and the radiation optical field is *almost monochromatic* with its frequency being coincident with one of the frequencies *of atomic transitions* ($\simeq \omega_0 \rangle 0$). Then the *influence* of other atomic levels can be neglected. Thus, for each atom there remains only a *two-dimensional Hilbert space of state functions* $|+\rangle$ and $|-\rangle$ (a two-level atom). In this basis, corresponding to the dipole representation, the atomic part of the total Hamiltonian

$$\hat{H} = \hat{H}_{\mathbf{k}} + \hat{H}_{af} \,,$$

where

$$\langle \pm | \hat{H}_{af} | \pm \rangle = E_{\pm}, \; \langle \pm | \hat{H}_{af} | \mp \rangle = 0 \,,$$

therefore, H_{af} may be written in the quasi-spin representation as

$$\hat{H}_{af} = \frac{E_+ + E_-}{2} \mathbb{I} + \frac{1}{2} \hbar \omega_0 \sigma_f^z$$

$$\sigma_f^z \, |\pm\rangle = \pm |\pm\rangle, \quad [\, \sigma_f^z, \sigma_\pm] = \pm 2\sigma_\pm \,,$$

where $f \in \mathcal{A}$ is the index of an atom in the system. Next, the electromagnetic field is quantized in the volume $V \geq V_c$ of the active medium:

$$\mathbf{A} = \sum_{\mathbf{q},\alpha} \left(\frac{2\pi\hbar c}{\omega_{\mathbf{q}} V} \right)^{\frac{1}{2}} \{\psi_{\mathbf{q}\alpha}(x) b_{\mathbf{q}\alpha} + b^+_{\mathbf{q}\alpha} \psi^*_{\mathbf{q}\alpha}(x)\} ,$$

$b^+_{\mathbf{q}\alpha}(b_{\mathbf{q}\alpha})$ are creation (annihilation or destruction) operators of a photon with frequency $\omega_{\mathbf{q}}$ and a complete set of functions $\psi_{\mathbf{q}\alpha}(x)$ in the volume V :

$$\psi_{\mathbf{q}\alpha}(x) = V^{-\frac{1}{2}} \mathbf{e}_{\mathbf{q}} \chi_\alpha e^{i\langle\, \mathbf{q},\mathbf{x}\,\rangle} ,$$

specifying the momentum of a photon and its polarization $\alpha : \langle\, \alpha, \mathbf{q}\,\rangle = \pm|\mathbf{q}|$. Hence,

$$\hat{H} = \underbrace{\sum_{\mathbf{q},\alpha} \hbar\omega_{\mathbf{q}} b^+_{\mathbf{q}\alpha} b_{\mathbf{q}\alpha}}_{\textit{photon part}} \bigoplus \underbrace{\frac{E_+ + E_-}{2} N\mathbb{I} + \sum_f \frac{1}{2}\hbar\omega_0 \sigma^z_f}_{\textit{electron part}} \bigoplus$$

$$\underbrace{\sum_{\alpha,f} \frac{g}{\sqrt{N}} (\sigma^-_f b^+_{0,\alpha} + \sigma^+_f b_{0,\alpha})}_{\textit{interaction part}} ,$$

where $\sigma_f \pm |0\,\rangle = |\pm\rangle$, and $N \in \mathbb{Z}_+$ is the amount of atoms in the medium. The parameter $g \in \mathbb{R}_+$, coupling photon and electron fields is expressed as $g = \sqrt{2\pi\hbar\omega_0\rho}$, with $\rho = \frac{N}{V}$ being the atom density. The above obtained Hamiltonian describes an equilibrium *phase transition* to the state of ordered electron dipoles that is sometimes called a *super-radiance* state and the process is known as that of spontaneous *coherent monochromatic radiation.*

Let us now perform the so called Bogoliubov canonical transformation that shifts the *bosonic variables*:

$$\tilde{b}^+_{0,\alpha} = b^+_{0,\alpha} + \frac{g}{\hbar\omega_0 N^{\frac{1}{2}}} \sum_f \sigma^+_f \,. \tag{14.1}$$

Due to (14.1) the Hamiltonian arrives at the following form:

$$\begin{aligned} \hat{H} &= \sum_{\substack{q\neq q_0 \\ \alpha\neq\alpha_0}} \hbar\omega_{\mathbf{q}} b^+_{\mathbf{q},\alpha} b_{\mathbf{q},\alpha} + \sum_\alpha \hbar\omega_0 \tilde{b}^+_{0,\alpha_0} \tilde{b}_{0,\alpha_0} \bigoplus \\ &\bigoplus \frac{N(E_+ + E_-)}{2} \mathbb{I} + \sum_f \frac{1}{2}\hbar\omega_0 \sigma^z_f - \frac{g^2}{\hbar\omega_0 N} \sum_{ff'} \sigma^+_f \sigma^-_{f'} \,. \end{aligned} \tag{14.2}$$

It may be shown that for $N \to \infty$ in (14.1) the last term *may be replaced by a c-number.* This can be done due to the phenomenon of boson condensation. Thus, we can *decouple* the spaces of states of bosons and quasispin subsystems and analyze the thermodynamic properties of each part of the Hamiltonian (2) separately.

Chapter 15

The Bloch-Maxwell Super-Radiance Quantum Optical Model

As is well known, the resistance in conductors is determined by the interaction of electrons with ion vibrations in the metal lattice. Thus interaction is described by the diagram:

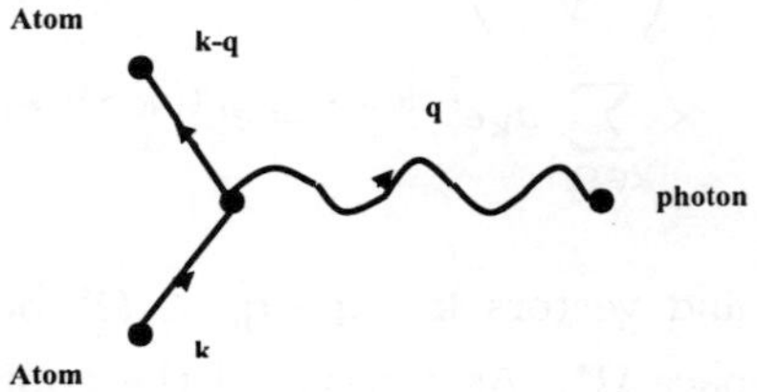

Within the dipole approximation, a two-level atomic gas in volume $\Omega \subset \mathbb{R}^3$interacting with electromagnetic field [Allen and Eberley, 1974] is usually described by the Hamiltonian.

$$\hat{H} = \frac{\omega_{12}}{2} \int\limits_{\Omega} d^3x \left[n(x) + \frac{N_0}{V} \right] + \sum_{\mathbf{q} \in \Omega^*} \omega_{\mathbf{q}} a_{\mathbf{q}}^+ a_{\mathbf{q}} -$$

$$-\delta \int\limits_{\Omega} d^3x \left[E^+(x)P(x) + P^+(x)E(x) \right], \tag{15.1}$$

where $\omega_{12} \in \mathbb{R}_+$ is the inter-level transition frequency, $\delta \in \mathbb{R}_+$ is the dipole moment of an atom in the gas. Operators $P : \mathcal{H} \to \mathcal{H}$ and $n : \mathcal{H} \to \mathcal{H}$ enjoying commutation relations

$$\begin{aligned} &[P^+(x), P(y)] = n(x)\delta(x-y), [P(x), n(y)] = 2P(x)\delta(x-y), \\ &[n(x), P^+(y)] = 2P^+(x)\delta(x-y), \ x, y \in \Omega, \end{aligned} \tag{15.2}$$

are operators of polarization and population of atoms in the gas. The electromagnetic field strength as an operator $E : \mathcal{H} \to \mathcal{H}$ can be expanded as follows:

$$E(x) = i \sum_{\mathbf{q} \in \Omega^*} \left(\frac{2\pi\omega_{\mathbf{q}}}{V} \right)^{\frac{1}{2}} a_{\mathbf{q}} e^{i<\mathbf{q},\mathbf{x}>} \tag{15.3}$$

where $\omega_{\mathbf{q}} := c|\mathbf{q}|$ is a photon frequency, c is the light velocity and $V = |\Omega|$ is the volume of a domain $\Omega \subset \mathbb{R}^3$. Within the circular field approximation the expression (15.3) can be restricted to the impact of only resonance photons:

$$E(x) \simeq i \left(\frac{2\pi\omega_{12}}{V} \right)^{\frac{1}{2}} \exp\left[i\left(< \mathbf{q}_0, \mathbf{x} > -\omega_{12} t\right)\right]$$

$$\times \sum_{\mathbf{k} \in \Omega^*} a_{\mathbf{k}} e^{i<\mathbf{k},\mathbf{x}>} = e^{[i(\langle \mathbf{q}_0, \mathbf{x}\rangle - \omega_{12} t)]} \varepsilon(x) \tag{15.4}$$

where $|\mathbf{q}_0| = \dfrac{\omega_{12}}{c}$ and vectors $\mathbf{k} = \mathbf{q} - \mathbf{q}_0 \in \Omega^*$ belong to the Fourier-conjugated vector space Ω^*. As a result of the approximation (15.4) the following commutation relation holds:

$$[\varepsilon(x), \varepsilon^+(y)] = 2\pi\omega_{12}\delta(x - y) \tag{15.5}$$

for all $x, y \in \Omega$, $t \in \mathbb{R}$. Similarly one can extract the fast oscillating part of the polarization operator $P : \mathcal{H} \to \mathcal{H}$:

$$P(x) = e^{[i(<\mathbf{q}_0,\mathbf{x}> - \omega_{12} t)]} p(x) \tag{15.6}$$

for $x \in \Omega$, $t \in \mathbb{R}$. Having as usual scaled the physical units so that $\hbar = c = \dfrac{l}{N_0} = 1$, where $l \in \mathbb{R}_+$ is the length of a gas domain being chosen such that $l >> \sqrt{S}$, S is its cross-section area, and $N_0 \in \mathbb{Z}_+$ is the amount of atoms in volume $\Omega \subset \mathbb{R}^3$, our Hamiltonian (15.1) takes, within the Wiegman-Rupasov approximation [Rupasov, 1982], the form:

$$\hat{H} = -i \int_{\mathbb{R}} dx \varepsilon^+(x) \varepsilon_x(x) - \varkappa \int_{\mathbb{R}} dx [\varepsilon^+(x) p(x) + p^+(x) \varepsilon(x)], \tag{15.7}$$

where the interaction constant $\varkappa = \left(\frac{2\pi\omega_{12}\delta^2}{S}\right)^{\frac{1}{2}}$ and the following commutation relations hold:

$$\begin{aligned}
&[\varepsilon(x), \varepsilon^+(y)] = \delta(x-y), \quad [p(x), n(y)] = 2p(x)\delta(x-y)\,, \\
&[p^+(x), p(y)] = n(x)\delta(x-y), \quad [n(x), p^+(y)] = 2p^+(x)\delta(x-y)\,, \\
&[n(x), n(y)] = 0 \quad = [n(x), \varepsilon(y)]
\end{aligned} \tag{15.8}$$

for all $x, y \in \Omega$. In the Hamiltonian (15.7) we have omited the insignificant term

$$\omega_{12}\int_{\mathbb{R}} dx \left(\varepsilon^+(x)\varepsilon(x) + \frac{1}{2}[n(x)+1]\right), \tag{15.9}$$

commuting with (15.7). This simply means that our new energy-momentum reference frame is put at the intersection point of energy and momentum spectra of the corresponding free fields.

Making use now of (15.7) and (15.8) one gets the following quantum dynamical system of operator fields:

$$\begin{aligned}
&\frac{d\varepsilon}{dt} = -\varepsilon_x + i\varkappa p, \quad \frac{dp}{dt} = -i\varkappa\varepsilon n\,, \\
&\frac{dn}{dt} = i2\varkappa(p^+\varepsilon - \varepsilon^+ p)\,,
\end{aligned} \tag{15.10}$$

where $t \in \mathbb{R}$ is the evolution parameter.

As one sees from (15.10), the dynamics of all physical quantities in (15.7) are strictly nonlinear; an interesting quantum optical model that is highly important for applications. We shall discuss this in more detail below, making use of results due to the so called quantum inverse spectral transform method devised recently especially for such spatially one-dimensional models.

Let us introduce a two-component fermi-field $\{\psi_\nu(x) : \nu = \overline{1,2},\ x \in \mathbb{R}\}$, enjoying the following commutation relations:

$$\begin{aligned}
&[\psi_\alpha(x), \psi_\beta^+(y)]_+ = \delta_{\alpha\beta}\delta(x-y)\,, \\
&[\psi_\alpha(x), \psi_\beta(y)]_+ = 0 = [\psi_\alpha^+(x), \psi_\beta^+)]_+
\end{aligned} \tag{15.11}$$

for all $\alpha, \beta = \overline{1,2}$ and $x, y \in \mathbb{R}$, being connected with quantities $p, n : \mathcal{H} \to \mathcal{H}$:

$$\begin{aligned} n(x) &= \psi_2^+(x)\psi_2(x) - \psi_1^2(x)\psi_1(x)\,, \\ p(x) &= \psi_1^+(x)\psi_2(x)\,. \end{aligned} \tag{15.12}$$

Having involved still on the ψ-spinor the completeness condition

$$\psi_1^+(x)\psi_1(x) + \psi_2^+(x)\psi_2(x) = 1\,, \tag{15.13}$$

one can satisfy all of the commutation relations (15.8). As a result, the operators of the number of particles and energy take the form:

$$\hat{N} = \int_{\mathbb{R}} dx[\varepsilon^+(x)\varepsilon(x) + \psi_2^+(x)\psi_2(x)]\,, \tag{15.14}$$

$$\hat{H} = -i\int_{\mathbb{R}} dx\varepsilon^+(x)\varepsilon_x(x) - \varkappa\int_{\mathbb{R}} dx[\psi_1^+(x)\varepsilon^+(x)\psi_2(x) + \psi_2^+(x)\varepsilon(x)\psi_1(x)]\,.$$

We now proceed to describing the eigenstates of the Hamiltonian operators (15.14) in the corresponding Fock space Φ. Denoting, as usual the vacuum state containing no photon and no excited gas atome by $|0> \in \Phi$. Its properties look as follows:

$$\varepsilon(x)|0\rangle = 0, \quad \psi_2(x)|0\rangle = 0, \quad \psi_1(x)|0\rangle \neq 0 \tag{15.15}$$

for all $x \in \mathbb{R}$. Assume now that the state $|\mu> \in \Phi$ possesses a one excited atom with the total energy $-\mu \in \mathbb{R}$:

$$\hat{N}|\mu\rangle = |\mu>, \quad \hat{H}|\mu\rangle = -\mu|\mu\rangle\,. \tag{15.16}$$

In terms of our local operator fields the state $|\mu> \in \Phi$ can be represented as

$$|\mu\rangle = i\varkappa\int_{\mathbb{R}} dx e^{ikx}\left[\varepsilon^+(x) + \frac{\varkappa}{\mu\psi_2^+(x)\psi_1(x)}\right]|0\rangle\,, \tag{15.17}$$

where $k = \dfrac{\varkappa^2}{\mu} - \mu$ means the quasi-momentum of our excited quasiparticle. The energy spectrum of the state $|\mu\rangle \in \Phi$ consists of two polariton [Allen and Eberly, 1974] branches:

$$\omega_\pm(k) = \frac{k}{2} \pm \left(\frac{k^2}{4} + \varkappa^2\right)^{\frac{1}{2}}\,, \tag{15.18}$$

where as before, $k \in \mathbb{R}$ is the quasiparticle momentum. Similarly one can construct many-quasiparticle states $|\mu_1, \mu_2, \ldots, \mu_N\rangle \in \Phi$ with arbitrary $N \in \mathbb{Z}_+$, enjoing the following properties:

$$\begin{aligned} \hat{N}|\mu_1, \mu_2, \ldots, \mu_N\rangle &= N|\mu_1, \mu_2, \ldots, \mu_N\rangle\,, \\ \hat{H}|\mu_1, \mu_2, \ldots, \mu_N\rangle &= -\sum_{j=1}^{N} \mu_j |\mu_1, \mu_2, \ldots, \mu_N\rangle\,. \end{aligned} \tag{15.19}$$

Here $\mu_j \in \mathbb{C}$, $j = \overline{1, N}$, can be chosen arbitrarily. These states describe $N \in \mathbb{Z}_+$ free quasi-particles existing in the model if $\sum\limits_{j=1}^{N} \mu_j$ is real. Let us now consider the case when energy parameters $\mu_j \in \mathbb{C}\backslash\mathbb{R}$, $j = \overline{1, N}$, where by definition

$$\mu_j = \frac{\mu}{N} + i\varkappa^2 \left(\frac{N+1}{2} - j \right), \; \mu = \sum_{j=1}^{N} \mu_j\,, \tag{15.20}$$

and the total energy $-\mu \in \mathbb{R}_-$ of the corresonding state $|(\mu)_N\rangle \in \Phi$ will be real and negative at $\mu \in \mathbb{R}_+$:

$$\hat{H}|(\mu)_N\rangle = -\mu|(\mu)_N\rangle, \quad \hat{N}|(\mu)_N\rangle = N|(\mu)_N\rangle\,. \tag{15.21}$$

Therely we can interprete the state $|(\mu)_N\rangle \in \Phi$ as a bound state of $N \in \mathbb{Z}_+$ tied quasiparticles, being still often called a quantum N-soliton.

The bound quasi-particle states described above and called quantum solitons apart of usual free quasiparticle states, possess the spatial size $r_0 > 0$ being reciprocal to a number of bound particles. The latter makes it possible to interprete them as the proper super-radiance impulses generated by the physical quantum optical system modeled above by the Bloch-Maxwell Hamiltonian system (15.10).

Chapter 16

A Degenerate Parametric Quantum Optical Amplifier

One of the simplest interactions in Nonlinear Optics is where a photon of frequency 2ω splits into two photons each with frequency ω. This process is known as *parametric down conversion* to describe a *parametric amplifier.* In a parametric amplifier a signal at frequency ω is *amplified* by pumping a crystal with χ^2-nonlinearity at frequency 2ω. We consider a simple model where *the pump* mode at frequency 2ω *is classical* and the signal mode at frequency ω is described by the creation annihilation operators a^+, a of a photon. Then the corresponding Hamiltonian is

$$\hat{H} = \hbar\omega a^+ a - i\hbar\frac{\chi}{2}(a^2 e^{2i\omega t} - a^{+2}e^{-2i\omega t}), \tag{16.1}$$

where χ is a constant proportional to the second order nonlinear susceptibility. The first step is to pass to the interaction picture:

$$\begin{aligned} \hat{H}_0 &= \hbar\omega a^+ a, \\ a &:\longrightarrow \exp\left(-\frac{i}{\hbar}\hat{H}_0 t\right) a \exp\left(\frac{i}{\hbar}\hat{H}_0 t\right) \\ a_I &= \exp\left(-\frac{i}{\hbar}\hbar\omega a^+ at\right) a \exp\left(\frac{i}{\hbar}\hbar\omega a^+ at\right) \\ &= \exp(-i\omega a^+ at) a \exp(i\omega a^+ at) \\ &= a\exp(i\omega t) \Longrightarrow a = a_I = \exp(-i\omega t) \end{aligned} \tag{16.2}$$

As a result we can obtain the Hamiltonian operator in the interaction representation (picture) as follows:

$$\tilde{H} = \hbar\omega a_I^+ a_I + i\hbar\frac{\chi}{2}(a_I^{+2} - a_I^2). \tag{16.3}$$

Now our first task is to find the spectrum of (16.3) in the corresponding Hilbert Fock type space as one sees that $\tilde{H}$ is time-independent. To do this it is necessary to diagonalize (16.3) in some appropriate representation of quasi-particle states. The canonical transformation is due to:

$$\begin{aligned} a_I &= u\tilde{a} + v\,\tilde{a}^+, \\ a_I^+ &= u^*\tilde{a}^+ + v^*\,\tilde{a}\,, \end{aligned} \tag{16.4}$$

where $|u|^2 - |v|^2 = 1$, and operators $\tilde{a}$, $\tilde{a}^+ \colon \Phi \to \Phi$ satisfy the standard Bose-commutation relationships:

$$[\tilde{a}, \tilde{a}^+] = 1\,. \tag{16.5}$$

Having substituted (16.4) into (16.3), one gets

$$\begin{aligned} \tilde{H} &= \hbar\omega(|u|^2\tilde{a}^+\tilde{a} + |v|^2\tilde{a}\tilde{a}^+) + \hbar\omega(u^*v\tilde{a}^{+2} + v^*u\tilde{a}^2) \\ &\quad + i\hbar\frac{\chi}{2}[u^{*2}\tilde{a}^{+2} + v^{*2}\tilde{a}^2 + u^*v^*(\tilde{a}^+\tilde{a} + \tilde{a}\tilde{a}^+)] \\ &\quad - i\hbar\frac{\chi}{2}[u^2\tilde{a}^2 + v^2\tilde{a}^{+2} + uv(\tilde{a}^+\tilde{a} + \tilde{a}\tilde{a}^+)] \\ &= \hbar\omega(|u|^2\tilde{a}^+\tilde{a} + |v|^2\tilde{a}^+\tilde{a} + |v|^2\mathbb{I}) + i\hbar\frac{\chi}{2}(u^*v^* - uv)(2\tilde{a}^+\tilde{a} + \mathbb{I}) \\ &\quad + \tilde{a}^{+2}\left(\hbar\omega u^*v + i\hbar\frac{\chi}{2}u^{*2} - i\hbar\frac{\chi}{2}v^2\right) \\ &\quad + \tilde{a}^2\left(\hbar\omega v^*u + i\hbar\frac{\chi}{2}v^{*2} - i\hbar\frac{\chi}{2}u^2\right). \end{aligned} \tag{16.6}$$

It is easy to see that the expression (16.6) is diagonalizable if the conditions

$$\begin{aligned} \hbar\omega u^*v + i\hbar\frac{\chi}{2}u^{*2} - i\hbar\frac{\chi}{2}v^2 = 0\,, \\ \hbar\omega v^*u + i\hbar\frac{\chi}{2}v^{*2} - i\hbar\frac{\chi}{2}u^2 = 0\,, \end{aligned} \tag{16.7}$$

hold for any real susceptibility $\chi \in \mathbb{R}_+$. From (16.7) one finds that

$$\begin{aligned} i\hbar\frac{\chi}{2} &= \hbar\omega\frac{u^*v}{v^2 - u^{*2}}\,, \\ i\hbar\frac{\chi}{2} &= \hbar\omega\frac{v^*u}{u^2 - v^{*2}}\,. \end{aligned} \tag{16.8}$$

Since $|u|^2 - |v|^2 = 1$, one can take

$$u = \cosh\varphi e^{i\alpha}, \qquad v = \sinh\varphi e^{i\beta}\,, \tag{16.9}$$

where $\varphi \in \mathbb{R}$ and $\alpha, \beta \in [0, 2\pi)$ some still undefined parameters. By substituting (16.9) into (16.8) we get that

$$i\frac{\chi}{2} = \frac{\omega \exp[i(\alpha - \beta)] \cosh\varphi \sinh\varphi}{\cosh^2\varphi e^{2i\alpha} - \sinh^2\varphi e^{-2i\beta}} = \frac{\omega \exp[i(\alpha - \beta)] \sinh 2\varphi}{2\left[\cosh^2\varphi - \sinh^2\varphi e^{-2i(\beta+\alpha)}\right] e^{2i\alpha}}$$

$$= \frac{\omega \sinh 2\varphi \left\{\cos(\alpha+\beta) - i\left(1 + 2\sinh^2\varphi\right)\sin(\alpha+\beta)\right\}}{2\left\{\left[\cosh^2\varphi - \sinh^2\varphi\cos 2(\alpha+\beta)\right]^2 + \sinh^4\varphi \sin^2 2(\alpha+\beta)\right\}}$$

$$= \frac{\omega \sinh 2\varphi \left\{\cos(\alpha+\beta) - i(1 + 2\sinh^2\varphi)\sin(\alpha+\beta)\right\}}{2\left[1 + \sinh^2\psi \sin^2 2(\alpha+\beta)\right]}. \tag{16.10}$$

Since the left hand side of (16.10) is purely imaginary, one easily finds that

$$\omega \sinh 2\varphi \cos(\alpha+\beta) \equiv 0 \ at \ \sinh 2\varphi \neq 0 \tag{16.11}$$

as $\chi \neq 0$. The condition (16.11) amounts to $\cos(\alpha+\beta) = 0$ or $(\alpha+\beta) = \frac{\pi}{2}$ or $\frac{3\pi}{2}, \frac{5\pi}{2}, \frac{7\pi}{2}$. Thereby we arrive at the following: at $\alpha = 0$, $\beta = \frac{3\pi}{2}$

$$a_I = \cosh\varphi\tilde{a} - i\sinh\varphi\,\tilde{a}^+,$$

$$a_I^+ = \cosh\varphi\tilde{a}^+ + i\sinh\varphi\,\tilde{a},$$

and the expression (16.10) with measures like $\sin(\alpha+\beta) = -1$

$$i\frac{\chi}{2} = \frac{\omega \sinh 2\varphi\left(1 + 2\sinh^2\varphi\right) i}{2\left\{\left(\cosh^2\varphi + \sinh^2\varphi\right)^2 + \sinh^4\varphi \cdot 0\right\}} = i\frac{\omega \sinh 2\varphi}{2\left(1 + 2\sinh^2\varphi\right)} \tag{16.12}$$

The result (16.12) clearly means that

$$\chi = \frac{\omega \sinh 2\varphi}{(1 + 2\sinh^2\varphi)} \tag{16.13}$$

defining uniquely the phase parameter $\varphi \in \mathbb{R}$:

$$\chi = \omega \tanh 2\varphi \tag{16.14}$$

Now, going back to the Hamiltonian expression (16.6):

$$\frac{1}{2}\tilde{H} = \hbar\omega\left[\cosh^2\varphi\tilde{a}^+\tilde{a} + \sinh^2\varphi(\tilde{a}^+\tilde{a} + \mathbb{I})\right]$$

$$+ i\hbar\frac{\chi}{2}\left(\sinh 2\varphi\frac{i}{2} + \sinh 2\varphi\frac{i}{2}\right)(2\tilde{a}^+\tilde{a} + \mathbb{I})$$

$$
\begin{aligned}
&= \hbar\omega\tilde{a}^{+}\tilde{a}\left\{\cosh^2\varphi + \sinh^2\varphi\right\} + \hbar\omega\sinh^2\varphi\mathbb{I} \\
&\quad + (-1)(2\tilde{a}^{+}\tilde{a} + \mathbb{I})\sinh 2\varphi\frac{\hbar\chi}{2} \\
&= [\hbar\omega\left(1 + 2\sinh^2\varphi\right) - \chi\hbar\sinh 2\varphi]\tilde{a}^{+}\tilde{a} \\
&\quad + \left(\hbar\omega\sinh^2\varphi - \sinh 2\varphi\frac{\hbar\chi}{2}\right)\mathbb{I} \\
&= \frac{\hbar\omega}{\cosh 2\varphi}\tilde{a}^{+}\tilde{a} - \hbar\omega\frac{\sinh^2\varphi}{\cosh 2\varphi}\mathbb{I}\,.
\end{aligned}
\tag{16.15}
$$

The result (16.15) means that due to the interaction with pumping medium the effective photon zero energy level is lowered by the value $-\hbar\omega\dfrac{\sinh^2\varphi}{\cosh 2\varphi} = \triangle E_0 < 0$. This energy can obviously be used to create a real photon held in the vacuum state $|\tilde{0}\,\rangle \in \Phi$ of the effective Hamiltonian (16.15), equal to

$$
\frac{1}{2}\tilde{H} = \frac{\hbar\omega}{\cosh 2\varphi}\tilde{a}^{+}\tilde{a} - \hbar\omega\frac{\sinh^2\varphi}{\cosh 2\varphi}\mathbb{I}\,. \tag{16.16}
$$

To show that this process can really be realized, let us calculate the average number of created photons in the vacuum state $|\tilde{0}\,\rangle$ of the Hamiltonian (16.16):

$$
\begin{aligned}
N_I^{(0)} &= \langle\,\tilde{0}|a_I^{+}a_I|\tilde{0}\,\rangle = \langle\,\tilde{0}|(\cosh\varphi - \sinh\varphi)(\cosh\varphi - \sinh\varphi)|\tilde{0}\,\rangle \\
&= \sinh^2\varphi\langle\,\tilde{0}|\tilde{a}\tilde{a}^{+}|\tilde{0}\,\rangle = \sinh^2\varphi\,.
\end{aligned}
\tag{16.17}
$$

For the creation process to be real, we clearly need to have $N_I \geq 1$. Recalling that due to (16.14) $\chi = \omega\tanh 2\varphi$, one gets that

$$
\begin{aligned}
\frac{\chi}{\omega} &= \frac{\sinh 2\varphi}{\cosh 2\varphi} = \frac{2\sinh\varphi\cosh\varphi}{1 + 2\sinh^2\varphi} \\
&= 2\frac{\sqrt{\sinh^2\varphi}\sqrt{1 + \sinh^2\varphi}}{1 + 2\sinh^2\varphi} \cong 2\frac{\sqrt{N_I(1 + N_I)}}{1 + 2N_I^2}\,.
\end{aligned}
\tag{16.18}
$$

From (16.18) and with $N_I = 1$, one finds that

$$
\frac{\chi}{\omega} \simeq \frac{2\sqrt{2}}{3} \simeq 0.9(!)\,. \tag{16.19}
$$

If $N_I \geq 1$, we see that it makes sense to excite the next energy level $|\tilde{1}\,\rangle = \tilde{a}^{+}|\tilde{0}\,\rangle$ of the Hamiltonian (16.16). Then the average number of real photons in the system will be equal to the amount

$$N_I^{(1)} = \langle\, \tilde{1}|a_I^+ a_I|\tilde{1}\,\rangle = \cosh^2\varphi + \sinh^2\varphi = 1 + 2\sinh^2\varphi = \cosh 2\varphi\,\rangle 1 \tag{16.20}$$

Similarly to (16.17) the result can be obtained when finding the average number of photons created due to the interaction in the old vacuum state $|0_I\,\rangle \in \Phi$:

$$\begin{aligned}\tilde{N}_I &= \langle\, 0_I|\tilde{a}^+\tilde{a}|0_I\,\rangle \\ &= \langle\, 0_I|(\cosh\varphi a_I^+ - i\sinh\varphi a_I) + (\cosh\varphi a_I + i\sinh\varphi a_I^+)|0_I\,\rangle \\ &= \sinh^2\varphi\,.\end{aligned} \tag{16.21}$$

where we have used the expressions stemmed from (16.4):

$$\tilde{a} = i\sinh\varphi a_I^+ + \cosh\varphi a_I,\ \tilde{a}^+ = \cosh\varphi a_I^+ - i\sinh\varphi a_I\,.$$

We found that the vacuum state $|0_I\,\rangle \in \Phi$ contains (due to the interaction of the medium with energy pumping source) the same amount of photons as there are in the vacuum state $|\tilde{0}\,\rangle \in \Phi : N_I = \tilde{N}_I = \sinh^2\varphi > 0$. Hence the *active crystal medium* really realizes the *process of parametric down conversion* describing *a parametric amplifier.* Due to the fact that $a_I = \exp(i\omega t)a$, $a_I^+ = \exp(-i\omega t)a^+$ due to the initially made operator transformation, we see that the effective photon operator quantities *are squeezed* due to the interaction: if $\tilde{X} := \tilde{a} + \tilde{a}^+$, $\tilde{Y} = \dfrac{(\tilde{a}^+ - \tilde{a})}{i}$ - real measurable quantities we see that at $X = a_I^+ + a_I$, $Y = \dfrac{(a_I^+ - a_I)}{i}$

$$\tilde{X} = \cosh\varphi X - \sinh\varphi Y, \quad \tilde{Y} = \cosh\varphi Y - \sinh\varphi X\,. \tag{16.22}$$

It is interesting to find also the average number of photons in coherent states $|\tilde{\alpha}\,\rangle \in \Phi$ prepared from effective quasi-photon particles in the Fock space Φ :

$$|\tilde{\alpha}\,\rangle = \exp(\alpha\tilde{a}^+ - \alpha^*\tilde{a})|\tilde{0}\,\rangle \in \Phi\,, \tag{16.23}$$

where $\alpha \in \mathbb{C}$ is arbitrary complex parameter. The operator

$$\tilde{D}(\alpha) = \exp(\alpha\tilde{a}^+ - \alpha^*\tilde{a}), \quad \tilde{a}|\tilde{\alpha}\,\rangle = \alpha|\tilde{\alpha}\,\rangle\,, \tag{16.24}$$

is evidently unitary, $\tilde{D}(\alpha)\tilde{D}^*(\alpha) = 1$, rendering the normalizing condition of the state (16.23):

$$\langle\, \tilde{\alpha}|\tilde{\alpha}\,\rangle = 1,\ \tilde{D}^{-1}(\alpha)a\tilde{D}(\alpha) = \tilde{a} + \alpha\,, \tag{16.25}$$

for all $\alpha \in \mathbb{C}$. Our task now is to compute the average number of real photons over the state $\langle\, \tilde{\alpha}| \in \Phi$, *randomly distributed*:

$$\begin{aligned}
\tilde{N}_I &= \int\limits_{\mathbb{C}^2} \langle\, \tilde{\alpha}|a_I^+ a_I|\tilde{\alpha}\,\rangle P_N(\alpha) \left(\frac{d\alpha^* \wedge d\alpha}{2i}\right) \\
&= Tr \int\limits_{\mathbb{C}^2} (a_I^+ a_I)|\tilde{\alpha}\,\rangle\langle\, \tilde{\alpha}| \left(\frac{d\alpha^* \wedge d\alpha}{2i}\right) P_N(\alpha) \\
&= Tr(a_I^+ a_I \tilde{\rho}_N(\alpha))\,, \qquad (16.26)
\end{aligned}$$

where, by definition,

$$\tilde{\rho}_N(\alpha) = |\tilde{\alpha}\,\rangle\langle\, \tilde{\alpha}|P_N(\alpha), \quad P_N(\alpha) = \frac{1}{\pi N}\exp(-|\alpha|^2/N)\,. \qquad (16.27)$$

As a result, we can find that

$$\begin{aligned}
\tilde{N}_I &= Tr(a_I^+ a_I \tilde{\rho}_N(\alpha)) \Longrightarrow \int\limits_{\mathbb{C}^2} \frac{d\alpha^* \wedge d\alpha}{2i} P_N(\alpha)\langle\, \tilde{\alpha}|a_I^+ a_I|\tilde{\alpha}\,\rangle \\
&= \int\limits_{\mathbb{C}^2} \frac{d\alpha^* \wedge d\alpha}{2i} P_N(\alpha)\langle\, \tilde{\alpha}|(\cosh^2\varphi \tilde{a}^+\tilde{a} + \sinh^2\varphi \tilde{a}\tilde{a}^+ \\
&\quad + \cosh\varphi \sinh\varphi i\tilde{a}^2 - i\cosh\varphi\sinh\varphi\tilde{a}^{+2})|\tilde{\alpha}\,\rangle \\
&= \int\limits_{\mathbb{C}^2} \frac{d\alpha^* \wedge d\alpha}{2i} P_N(\alpha)\langle\, \tilde{\alpha}|\cosh^2\varphi\tilde{a}^+\tilde{a}|\tilde{\alpha}\,\rangle \\
&\quad + \int\limits_{\mathbb{C}^2} \frac{d\alpha^* \wedge d\alpha}{2i} P_N(\alpha)\langle\, \tilde{\alpha}|(\sinh^2\varphi\tilde{a}^+\tilde{a} + 1)|\tilde{\alpha}\,\rangle \\
&\quad + i\cosh\varphi\sinh\varphi \int\limits_{\mathbb{C}^2} \frac{d\alpha^* \wedge d\alpha}{2i} P_N(\alpha)\langle\, \tilde{\alpha}|\tilde{a}^2|\tilde{\alpha}\,\rangle \\
&\quad - i\cosh\varphi\sinh\varphi \int\limits_{\mathbb{C}^2} \frac{d\alpha^* \wedge d\alpha}{2i} P_N(\alpha)\langle\, \tilde{\alpha}|\tilde{a}^{+2}|\tilde{\alpha}\,\rangle \\
&= \int\limits_{\mathbb{C}^2} \frac{d\alpha^* \wedge d\alpha}{2i} P_N(\alpha)\alpha^*\alpha(\cosh^2\varphi + \sinh^2\varphi)
\end{aligned}$$

$$+ \sinh^2 \varphi \int_{\mathbb{C}^2} \frac{d\alpha^* \wedge d\alpha}{2i} P_N(\alpha)$$

$$+ i \sinh \varphi \cosh \varphi (\int_{\mathbb{C}^2} \frac{d\alpha^* \wedge d\alpha}{2i} P_N(\alpha)(\alpha^2 - \alpha^{*2}))$$

$$= \cosh 2\varphi N + \sinh^2 \varphi + \frac{\sinh 2\varphi}{2} \left(-\frac{4N^2}{\pi 4N} \right)$$

$$= \left(\cosh 2\varphi - \sinh \frac{2\varphi}{2\pi} \right) N + \sinh^2 \varphi$$

$$= \left(\cosh 2\varphi - \sinh \frac{2\varphi}{2\pi} \right) N + \bar{N}_I \tag{16.28}$$

We can explain the result (16.28) as the re-distribution of $N \in \mathbb{Z}_+$ previously generated (quasi) photons within the coherently prepared quasi-states $\langle\, \tilde{\alpha}| \in \Phi$ where an additional photon particle is added due to the term $\bar{N}_I = \sinh^2 \varphi$ on the right hand side of (16.28). This confirms once more that our interacting crystal medium can be interpreted as *based on parametric down conversion* process *amplifier* of coherent radiation.

Note 1. Some motivations arguing the realization of condition (16.19) are based on the following facts: the true Hamiltonian describing a degenerate parametric amplifier via the parametric down-conversion process look like

$$\hat{H}_{int} = \hbar g (a_s^+ a_i^+ b + a_s a_i b^+) \,, \tag{16.29}$$

(g-the coupling constant) where a_s, a_i–are annihilation operators of the signal and idler-modes correspondingly, b is the annihilation operator of the pumping source, interacting with the medium. As often, the pump depletion is neglected. Then the Hamiltonian (16.29) becomes

$$\hat{H}_{int} = \hbar g \beta_p (a_s^+ a_i^+ \exp(-i\varphi) - a_s a_i \exp(i\varphi)) \,, \tag{16.30}$$

where β_p and $\varphi = 2\omega t$ are the real amplitude and phase of the coherent pump field. This approximation is valid in the limits

$$gt \to 0, \;\; \beta_p \to \infty \;\, at \;\; g\beta_p t = constant \,, \tag{16.31}$$

so the condition (16.19) is realizable.

$$
2\omega \xrightarrow{}
\boxed{\begin{array}{c} nonlinear \\ medium \end{array}}
\begin{array}{l} \nearrow \;\omega_1 \\[4pt] \searrow \;\omega_2 \end{array}
$$

Note 2. Another argument is true: the lowering of the zero level energy is $-\triangle E_0 = \dfrac{\sinh^2 \varphi}{\cosh 2\varphi}$, then this defect of energy produces $N = \dfrac{\triangle E_0}{\tilde{\omega}} = \sinh^2 \varphi$ particles in the vacuum approach to studying quantum nonlinear systems under nontrivial interactions.

Chapter 17

Two-Mode Nonlinear Quantum Optics System: A Non-degenerate Parametric Quantum Optical Amplifier

In two-mode systems there is a richer variety of quantum phenomena since there exists the possibility of quantum correlations between the modes. These correlations may give rise to two-mode squeezing of light states and so on.

The nondegenerate parametric (two-mode)-amplifier is interacting with light of frequency 2ω crystal pumping the energy into the system, described by two modes of light at frequencies ω_1 and ω_2, such that

$$2\omega = \omega_1 + \omega_2 .$$

We designate one mode as the *signal* and the other as the *idler.*

The Quantum Hamiltonian of this system is written down in the simplified form as

$$\hat{H} = \hbar\omega_1 a_1^+ a_1 + \hbar\omega_2 a_2^+ a_2 + i\hbar\chi(a_1^+ a_2^+ \exp(-2\omega t i) - a_1 a_2 \exp(2\omega t i)) , \quad (17.1)$$

where, bose-operators $a_j, a_k^+ : \Phi \to \Phi$, $j,k = \overline{1,2}$, satisfy the relationships

$$[a_j, a_k^+] = \delta_{jk} . \quad (17.2)$$

Making now use of the transformations

$$a_j \exp(it\omega_j) = a_{I,j}, \ a_j^+ \exp(-it\omega_j) = a_{I,j}^+ \quad (17.3)$$

for $j = \overline{1,2}$, one finds easily that

$$\hat{H}_I = \hbar\omega_1 a_{I,1}^+ a_{I,1} + \hbar\omega_2 a_{I,2}^+ a_{I,2} + i\hbar\chi(a_{I,1}^+ a_{I,2}^+ - a_{I,1} a_{I,2}) , \quad (17.4)$$

is a new effective Hamiltonian not depending explicitly on time. We are interested in stable quasi-photon states of the Hamiltonian (17.4) and inspecting the number of real photons in its vacuum state, thereby realizing the parametric down conversion process in a crystal interacting with light via (17.4).

Making now use of the transformations

$$\begin{aligned} a_{1,I} &= u_1\tilde{a}_1 + v_2\tilde{a}_2^+, \quad a_{2,I} = u_2\tilde{a}_2 + v_1\tilde{a}_1^+\,, \\ a_{1,I}^+ &= u_1^*\tilde{a}_1^+ + v_2^*\tilde{a}_2, \quad a_{2,I}^+ = u_2^*\tilde{a}_2^+ + v_1^*\tilde{a}_1\,, \end{aligned} \tag{17.5}$$

where

$$|u_1|^2 - |v_2|^2 = 1,\ |u_2|^2 - |v_1|^2 = 1,\ u_1v_1 - u_2v_2 = 0\,, \tag{17.6}$$

one can deduce (by substituting (17.5), (17.6) into (17.4)) the following expression:

$$\begin{aligned} \tilde{H} &= \{\hbar\omega_1|u_1|^2 + \hbar\omega_2|v_1|^2 + i\hbar\chi(u_1^*v_1^* - u_1v_1)\}\tilde{a}_1^+\tilde{a}_1 \\ &+ \{\hbar\omega_1|v_2|^2 + \hbar\omega_2|u_2|^2 + i\hbar\chi(u_2^*v_2^* - u_2v_2)\}\tilde{a}_2^+\tilde{a}_2 \\ &+ \{\hbar\omega_2|v_1|^2 + \hbar\omega_1|v_2|^2 + i\hbar\chi(u_2^*v_2^* - u_1v_1)\}\mathbb{I}\,, \end{aligned} \tag{17.7}$$

if the condition

$$\frac{\hbar\omega_1 u_1 v_2^* + \hbar\omega_2 v_1^* u_2}{u_1u_2 - v_1^*v_2^*} = i\hbar\chi \tag{17.8}$$

holds. For the condition (17.8) to be enjoyed, let us first resolve constraints (17.6). We have:

$$\begin{aligned} u_1 &= \cosh\varphi_1 e^{i\alpha_1}, \quad u_2 = \cosh\varphi_2 e^{i\alpha_2}\,, \\ v_2 &= \sinh\varphi_1 e^{i\beta_1}, \quad v_1 = \sinh\varphi_2 e^{i\beta_2}\,, \end{aligned} \tag{17.9}$$

under the constraint

$$\tanh\varphi_1 e^{i(\beta_1-\alpha_1)} = \tanh\varphi_2 e^{i(\beta_2-\alpha_2)}\,, \tag{17.10}$$

where $\alpha_j, \beta_j \in [0, 2\pi)$, $j = \overline{1,2}$. By substituting (17.9) and (17.10) into (17.7) one arrives at

$$\begin{aligned} \tilde{H} &= \{\hbar\omega_1\cosh^2\varphi_1 + \hbar\omega_2\sinh^2\varphi_2 + 2\hbar\chi\cosh\varphi_1\sinh\varphi_2\sin(\alpha_1+\beta_2)\}\tilde{a}_1^+\tilde{a}_1 \\ &+ \{\hbar\omega_1\sinh^2\varphi_1 + \hbar\omega_2\cosh^2\varphi_2 + 2\hbar\chi\cosh\varphi_2\sinh\varphi_1\sin(\beta_1+\alpha_2)\}\tilde{a}_2^+\tilde{a}_2 \\ &+ \{\hbar\omega_2\sinh^2\varphi_2 + \hbar\omega_1\sinh^2\varphi_1 + 2\hbar\chi\sinh\varphi_1\cosh\varphi_2\sin(\alpha_2+\beta_1)\}\mathbb{I}\,, \end{aligned} \tag{17.11}$$

where still the constraint (17.8) was not taken into account. For this to be done, we have:

$$\frac{(\hbar\omega_1 \sinh^2\varphi_1 + \hbar\omega_2 \sinh^2\varphi_2)e^{i(\alpha_1-\beta_1)}}{\cosh\varphi_1 \cosh\varphi_2[e^{i(\alpha_1+\alpha_2)} - \tanh\varphi_1 \tanh\varphi_2 e^{-i(\beta_1+\beta_2)}]\tanh\varphi_1} = i\hbar\chi\,, \tag{17.12}$$

or, dividing the two sides of (17.12) by '$\hbar$' and dropping on the left the factor $e^{i(\alpha_1-\beta_1)}$, one arrives at:

$$i\chi = \frac{(\omega_1 \sinh^2\varphi_1 + \omega_2 \sinh^2\varphi_2)}{\cosh\varphi_1 \cosh\varphi_2\{\exp[i(\alpha_2+\beta_1)] - \tanh\varphi_1 \tanh\varphi_2 \exp[-i(\alpha_1+\beta_2)]\}\tanh\varphi_1} \tag{17.13}$$

which is imaginary. This amounts to the condition that

$$\mathrm{Re}[\exp[i(\alpha_2+\beta_1)] - \tanh\varphi_1 \tanh\varphi_2 \exp[-i(\alpha_1+\beta_2)]] = 0\,. \tag{17.14}$$

From (17.14)

$$e^{i(\alpha_2+\beta_1)} - \tanh\varphi_1 \tanh\varphi_2 e^{-i(\alpha_1+\beta_2)} + e^{-i(\alpha_2+\beta_1)} - \tanh\varphi_1 \tanh\varphi_2 e^{i(\alpha_1+\beta_2)} \equiv 0$$

that is

$$\cos(\alpha_2+\beta_1) - \tanh\varphi_1 \tanh\varphi_2 \cos(\alpha_1+\beta_2) = 0 \tag{17.15}$$

On the other hand we must use the compatibility condition (4.10) in the form

$$\tanh\varphi_1 \exp[i(\alpha_2+\beta_1)] = \tanh\varphi_2 \exp[i(\alpha_1+\beta_2)] \tag{17.16}$$

by substituting it into (17.14):

$$\mathrm{Re}[\exp[i(\alpha_2+\beta_1)] - \tanh\varphi_1 \tanh\varphi_2 \frac{\tanh\varphi_2}{\tanh\varphi_1} \exp[-i(\alpha_2+\beta_1)]]$$
$$= \mathrm{Re}[\exp[i(\alpha_2+\beta_1)] - \tanh^2\varphi_2 \exp[-i(\alpha_2+\beta_1)]] \Longrightarrow 0 \tag{17.17}$$

As a result of (17.17),

$$\cos(\alpha_2+\beta_1) - \tanh^2\varphi_2 \cos(\alpha_2+\beta_1) = \cos(\alpha_2+\beta_1)[1-\tanh^2\varphi_2] = 0 \tag{17.18}$$

As $1-\tanh^2\varphi_2 \neq 0$, $(\alpha_2+\beta_1) \in \{\frac{\pi}{2}, \frac{3\pi}{2}, \frac{5\pi}{2}, \frac{7\pi}{2}\}$. On the other hand, from (17.10)

$$|\tanh\varphi_1| = |\tanh\varphi_2|\,, \tag{17.19}$$

there exist two possibilities:

$$\varphi_1 = \varphi_2 = \varphi \quad or \quad \varphi_1 = -\varphi_2 = \varphi\,, \tag{17.20}$$

where $\varphi \in \mathbb{R}^1$. For the first one

$$\textbf{I}\colon \alpha_2 + \beta_1 = \alpha_1 + \beta_2 (\text{mod} 2\pi) = \frac{3\pi}{2} \tag{17.21}$$

and for the second

$$\textbf{II}\colon \alpha_2 + \beta_1 = \alpha_1 + \beta_2 + \pi (\text{mod} 2\pi) = \frac{3\pi}{2} \tag{17.22}$$

Thus, from (17.13), at $(\alpha_2 + \beta_1) = \dfrac{3\pi}{2}$

$$\begin{aligned}
\frac{(\omega_1+\omega_2)\sinh^2\varphi}{\cosh^2\varphi\{-i-\tanh^2\varphi i\}\tanh\varphi} &= \frac{(\omega_1+\omega_2)\sinh{}^2\varphi}{\cosh^2\varphi\tanh\varphi(-i)[1+\tanh^2\varphi]} \\
&= \frac{(\omega_1+\omega_2)\tanh^2\varphi\cosh^2\varphi i}{\tanh\varphi\cosh 2\varphi} = \frac{(\omega_1+\omega_2)}{2}\frac{\sinh 2\varphi}{\cosh 2\varphi}i \\
&= \frac{(\omega_1+\omega_2)}{2}(\tanh 2\varphi)i = i\chi\,,
\end{aligned} \tag{17.23}$$

or simply

$$\chi = \frac{(\omega_1+\omega_2)}{2}(\tanh 2\varphi)\,. \tag{17.24}$$

Similarly if $\alpha_2 + \beta_1 = \alpha_1 + \beta_2 + \pi(\text{mod}2\pi) = \frac{3\pi}{2}$ (due to (17.22)),

$$\frac{\omega_1\sinh^2\varphi + \omega_2\sinh^2\varphi}{\cosh^2\varphi\{-i+\tanh^2\varphi(-i)\}} = \frac{(\omega_1+\omega_2)}{2}(\tanh 2\varphi)i = i\chi\,, \tag{17.25}$$

that is the same result as found in (17.24). Now we can proceed to calculating the Hamiltonian expression (17.11) for case **I**:

$$\begin{aligned}
\tilde{H}_{(\mathbf{I})} = &\left\{\hbar\omega_1\cosh^2\varphi + \hbar\omega_2\sinh^2\varphi + \hbar\frac{(\omega_1+\omega_2)}{2}\tanh 2\varphi\sinh 2\varphi(-1)\right\}\tilde{a}_1^+\tilde{a}_1 \\
&+ \left\{\hbar\omega_1\sinh^2\varphi + \hbar\omega_2\cosh^2\varphi - \hbar\frac{(\omega_1+\omega_2)}{2}\tanh 2\varphi\sinh 2\varphi\right\}\tilde{a}_2^+\tilde{a}_2 \\
&+ \left\{\hbar\omega_2\sinh^2\varphi + \hbar\omega_1\sinh^2\varphi - \hbar\frac{(\omega_1+\omega_2)}{2}\tanh 2\varphi\sinh 2\varphi\right\}\mathbb{I}
\end{aligned}$$

$$= \left\{ \begin{array}{l} \hbar\omega_1 \left(\cosh^2\varphi - \dfrac{\sinh 2\varphi \tanh 2\varphi}{2} \right) \\ \qquad + \hbar\omega_2 \left(\sinh^2\varphi - \dfrac{\sinh 2\varphi \tanh 2\varphi}{2} \right) \end{array} \right\} \tilde{a}_1^+ \tilde{a}_1$$

$$+ \left\{ \begin{array}{l} \hbar\omega_1 \left(\sinh^2\varphi - \dfrac{\sinh 2\varphi \tanh 2\varphi}{2} \right) \\ \qquad + \hbar\omega_2 \left(\cosh^2\varphi - \dfrac{\sinh 2\varphi \tanh 2\varphi}{2} \right) \end{array} \right\} \tilde{a}_2^+ \tilde{a}_2$$

$$+ \left\{ \hbar(\omega_1 + \omega_2) \left(\sinh^2\varphi - \frac{\sinh 2\varphi \tanh 2\varphi}{2} \right) \right\} \mathbb{I}$$

$$\Longrightarrow \left\{ \hbar\omega_1 \frac{\cosh^2\varphi}{\cosh 2\varphi} - \hbar\omega_2 \frac{\sinh^2\varphi}{\cosh 2\varphi} \right\} \tilde{a}_1^+ \tilde{a}_1$$

$$+ \left\{ -\hbar\omega_1 \frac{\sinh^2\varphi}{\cosh 2\varphi} + \hbar\omega_2 \frac{\cosh^2\varphi}{\cosh 2\varphi} \right\} \tilde{a}_2^+ \tilde{a}_2 - \hbar(\omega_1 + \omega_2) \frac{\sinh^2\varphi}{\cosh 2\varphi} \mathbb{I} . \tag{17.26}$$

For the energy of one particle states corresponding to the Hamiltonian (4.26) to be positive, it is necessary to enjoy the following conditions:

$$\hbar\omega_1 \frac{\cosh^2\varphi}{\cosh 2\varphi} - \hbar\omega_2 \frac{\sinh^2\varphi}{\cosh 2\varphi} > 0, \ \hbar\omega \frac{\cosh^2\varphi}{\cosh 2\varphi} - \hbar\omega_1 \frac{\sinh^2\varphi}{\cosh 2\varphi} > 0 \tag{17.27}$$

simultaneously; that is, if $\omega_1 = \omega_2 \tanh^2 \varphi k^2$, then

$$k^2 > 1 \ and \ \frac{\omega_1}{\omega_2} < \ k . \tag{17.28}$$

Thus, constraints (17.27) can be true iff $\dfrac{\omega_1}{\omega_2} < \ k$, or $\omega_1 \leq \omega_2 k$. The case $\omega_1 = \omega_2$ was discussed in chapter 16. Below, we proceed to considering case **II** . One has correspondingly that

$$\tilde{H}_{(\mathbf{II})} = \left\{ \hbar\omega_1 \cosh^2\varphi + \hbar\omega_2 \sinh^2\varphi - 2\pi \frac{(\omega_1 + \omega_2)}{2} \tanh 2\varphi \frac{\sinh 2\varphi}{2} \right\} \tilde{a}_1^+ \tilde{a}_1$$

$$+ \left\{ \hbar\omega_1 \sinh^2\varphi + \hbar\omega_2 \cosh^2\varphi - 2\pi \frac{(\omega_1 + \omega_2)}{2} \tanh 2\varphi \frac{\sinh 2\varphi}{2} (-1) \right\} \tilde{a}_2^+ \tilde{a}_2$$

$$+ \left\{ \hbar(\omega_1 + \omega_2) \left(\sinh^2\varphi - \frac{\sinh 2\varphi \tanh 2\varphi}{2} \right) \right\} \mathbb{I}$$

$$= \left\{ \hbar\omega_1 \frac{\cosh^2\varphi}{\cosh 2\varphi} - \hbar\omega_2 \frac{\sinh^2\varphi}{\cosh 2\varphi} \right\} \tilde{a}_1^+ \tilde{a}_1$$

$$+ \left\{ \hbar\omega_2 \frac{\cosh^2\varphi}{\cosh 2\varphi} - \hbar\omega_1 \frac{\sinh^2\varphi}{\cosh 2\varphi} \right\} \tilde{a}_2^+ \tilde{a}_2 - \hbar(\omega_1 + \omega_2) \frac{\sinh^2\varphi}{\cosh 2\varphi} \mathbb{I}, \tag{17.29}$$

exhibiting no difference between two cases **I** and **II**. As one can see from (17.29), the Hamiltonian (17.29) will be positive definite iff at the $\varphi \in \mathbb{R}_+^1$ found from (17.24) the energies of quasiparticles in eigenstates of (17.29) are positive, that is the conditions

$$\begin{aligned} \hbar\tilde{\omega}_1 &= \hbar\omega_1 \frac{\cosh^2\varphi}{\cosh 2\varphi} - \hbar\omega_2 \frac{\sinh^2\varphi}{\cosh 2\varphi} > 0 \\ \hbar\tilde{\omega}_2 &= \hbar\omega_2 \frac{\cosh^2\varphi}{\cosh 2\varphi} - \hbar\omega_1 \frac{\sinh^2\varphi}{\cosh 2\varphi} > 0 \end{aligned} \tag{17.30}$$

hold simultaneously. For (17.30) to be satisfied, let us define as before, a parameter $k > 0$ via the formula:

$$\frac{\omega_1}{\omega_2 \tanh^2\varphi} = k^2 . \tag{17.31}$$

Clearly then, $\tilde{\omega}_1 > 0$ if $k > 1$. Moreover, since $\tanh^2\varphi < 1$ for all $\varphi \in \mathbb{R}_+$, (17.31) yields

$$\frac{\omega_1}{\omega_2} < k^2 . \tag{17.32}$$

Further, for the frequency $\tilde{\omega}_2$ to be positive,

$$\begin{aligned} \tilde{\omega}_2 &= \omega_1 \frac{\cosh^2\varphi}{\cosh 2\varphi} - \omega_2 \frac{\sinh^2\varphi}{\cosh 2\varphi} = \frac{\omega_2^2 k^2 \sinh^2\varphi}{\omega_1 \cosh 2\varphi} - \frac{\omega_1 \sinh^2\varphi}{\cosh 2\varphi} \\ &= \frac{\omega_1 \sinh^2\varphi}{\cosh 2\varphi} \left(\frac{\omega_2^2 k^2}{\omega_1^2} - 1 \right) > 0 , \end{aligned} \tag{17.33}$$

or, equivalently, $\tilde{\omega}_2 > 0$ iff

$$k > \frac{\omega_1}{\omega_2} . \tag{17.34}$$

Since, $k^2 > k > 1$ for all $\varphi \in \mathbb{R}_+^1$ satisfying (17.24) and (17.31),the condition (17.34) is sufficient for the quasiparticle frequencies $\tilde{\omega}_1$ and $\tilde{\omega}_2$ to be positive. The latter is evidently equivalent to the stability of the vacuum state $|\tilde{0}\rangle \in \Phi$ of the Hamiltonian operator (17.26). As a physical consequence of the result obtained above one can expect the generation of photon particles from the vacuum state $|\tilde{0}\rangle \in \Phi$ due to the interaction with external radiation within the crystal under consideration. One easily finds, for instance, that

$$\begin{aligned} N_I^{(1)} &= \langle\, \tilde{0}|a_{1,I}^{+}a_{1,I}|\tilde{0}\,\rangle = \langle\, \tilde{0}|(u_1^*\tilde{a}_1^+ + v_2^*\tilde{a}_2)(u_1\tilde{a}_1 + v_2\tilde{a}_2^+)|\tilde{0}\,\rangle \\ &= (\langle\, \tilde{1}\tilde{0}|v_2^*)(v_2|\tilde{0}\tilde{1}\,\rangle) = |v_2|^2\langle\, \tilde{1}\tilde{0}|\tilde{0}\tilde{1}\,\rangle = |v_2|^2 = \sinh^2\varphi\,, \end{aligned} \tag{17.35}$$

where we have used that $\tilde{a}_j|\tilde{0}\,\rangle = 0$, $j = \overline{1,2}$. Similarly,

$$\begin{aligned} N_I^{(2)} &= \langle\, \tilde{0}|a_{2,I}^{+}a_{2,I}|\tilde{0}\,\rangle = \langle\, \tilde{0}|(u_2^*\tilde{a}_2^+ + v_1^*\tilde{a}_1)(u_2\tilde{a}_2 + v_1\tilde{a}_1^+)|\tilde{0}\,\rangle \\ &= (\langle\, \tilde{0}\tilde{1}|v_1^*)(v_1|\tilde{1}\tilde{0}\,\rangle) = |v_1|^2\langle\, \tilde{0}\tilde{1}|\tilde{1}\tilde{0}\,\rangle = |v_1|^2 = \sinh^2\varphi\,. \end{aligned} \tag{17.36}$$

Thereby, we see that a pumping beam at frequency 2ω due to the interaction with a nonlinear crystal produces two photon beams of frequencies ω_1 and ω_2 where $\omega_1 + \omega_2 = 2\omega$ from the vacuum state $|\tilde{0}\rangle \in \Phi$. As before, one can observe that the signal beam, here with frequency ω_1, is built from squeezed photon states this was verified many times experimentally. It is important to note here that two generated photon beams are strongly correlated and coherent.

Note. Some additional comments are still needed for the two mode parametric down-conversion process to be robust enough in experiment. As we saw before, our nonlinear crystal is interacting with a pumping energy beam of frequency ω_1 and ω_2 correspondingly. If we are interested in our experiment to detect only one photon beam signal of frequency ω_1 for instance, the other photon beam of frequency ω_2 should be depleted or stored within a crystal medium, making it *an idle* beam. (This was mentioned from the very beginning of Chapter 17). For this to be attained it is necessary to require that the quasiparticle spectrum $\tilde{\omega}_2$ be less than zero, that is $\tilde{\omega} < 0$ for the Hamilton (14.29). Then the Hamiltonian (17.29) with such a condition has no unitary equivalent bounded from below ground state $|\tilde{0}\,\rangle \in \Phi$; likewise, the energy spectrum for this Hamiltonian *is not bounded from below.* Physically, this also means that our photon-crystal system is subject to go over into the bose-condensation state aiming to fill in all states of negative energy and at last *to create* the stable ground state with positive defined energy of quasi-particles. This is often called "filling in Fermi sea" in a system of many particles. Concerning our Hamiltonian (17.29) $\tilde{\omega}_2 < 0$ is equivalent to the constraint $\omega_1/\omega_2 > k$. As we know that $\omega_1/\omega_2 < k^2$, one arrives at the following final condition:

$$k < \frac{\omega_1}{\omega_2} < k^2 \tag{17.37}$$

sufficient for an idle beam to be present in a crystal. Another interesting point of view on the parametric down-conversion process can be obtained from the fact that the transformed Hamiltonian $\tilde{H}$ (17.29) clearly possesses exactly two other operatorial conservation laws:

$$\hat{Q}_1 = \tilde{a}_1^+ \tilde{a}_1 \quad and \quad \hat{Q}_2 = \tilde{a}_2^+ \tilde{a}_2 \,, \tag{17.38}$$

which commute with each other too. Thus, one finds right away that for instance

$$\begin{aligned}
\hat{Q}_1 &= (u_2 a_1^+ - v_2^* a_2)(u_2^* a_1 - v_2 a_2^+) \\
&= |u_2|^2 a_1^+ a_1 + |v_2|^2 a_2 a_2^+ - (v_2^* u_2^* a_1 a_2 + u_2 v_2 a_1^+ a_2^+) \\
&= \cosh^2 \varphi a_1^+ a_1 + \sinh^2 \varphi (\mathbb{I} + a_2^+ a_2) - (\cosh\varphi \sinh\varphi \exp(-i(\alpha_2 + \beta_1)) a_1 a_2 \\
&\quad + \cosh\varphi \sinh\varphi \exp(i(\alpha_2 + \beta_1) a_1^+ a_2^+) \\
&= \cosh^2 \varphi a_1^+ a_1 + \sinh^2 \varphi a_2^+ a_2 + \sinh^2 \varphi \mathbb{I} - \frac{1}{2} \sinh 2\varphi (a_1 a_2 i - i a_1^+ a_2^+) \\
&= \cosh^2 \varphi a_1^+ a_2 + \sinh^2 \varphi a_2^+ a_2 + \sinh^2 \varphi \mathbb{I} \\
&\quad + \frac{i \sinh 2\varphi}{2} (a_1^+ a_2^+ - a_1 a_2) \,.
\end{aligned} \tag{17.39}$$

Since, $[\tilde{H}, \hat{Q}_1] = 0$, one sees that the operator

$$\hat{Q}_1 = \cosh^2 \varphi N_1 + \sinh^2 \varphi N_2 + \frac{i \sinh 2\varphi}{2} (a_1^+ a_2^+ - a_1 a_2)$$

counts the number of quasi particles in the system during the interaction.

To end this Chapter let us mention the inverse process only to that discussed above: the so called *parametric up-conversion* and described by the Hamiltonian:

$$\hat{H} = \hbar\omega_1 a_1^+ a_1 + \hbar\omega_2 a_2^+ a_2 + i\hbar\chi (a_1^+ a_2 \exp i(\omega_2 - \omega_1) - a_1 a_2^+ \exp i(\omega_1 - \omega_2)) \tag{17.40}$$

having another conservation law subject to *idler* and *signal* modes.

Chapter 18

A Nonlinear Quantum-Optical System at Equilibrium Radiative State

A lot of modern electronic devices are operating with optical signals generated due to the interaction of a working substance like crystal with an external laser radiation. As in most of these devices the interaction is nonlinear and processes making them stable are quantum, it is important to devise suitable approaches to studying their equilibrium radiative states subject to the external conditions imposed upon a system. This is the main aim of this chapter developing the modern quantum field theory methods for the analysis of radiative states in a nonlinear quantum-optical system having important applications in electronics. A great deal of studies was done concerning the problems of stability, bifurcation behavior and dynamical properties of atomic and molecular systems having applications in optical bi-stability of a laser cavity with a nonlinear crystal medium, in some micro-electronics and other devices based on nonlinear interaction processes with radiation. Subject to multi-photon excitations of poly-atomic molecules undergoing also a self-interaction via the Kerr effect the related processes can be modeled by means of the following quantum-optical approximated Hamiltonian operator

$$\hat{H} = \hbar\omega_0 a^+ a - \hbar\chi_0 a^+ a^2 a^+ + \chi(a^+ c + a c^+) + \hbar\omega_1 c^+ c\,, \tag{18.1}$$

acting in a Fock space Φ, where $\chi_0 \in \mathbb{R}_+$ and $\chi \in \mathbb{R}_+$ are coupling constant parameters, $a, a^+ : \Phi \to \Phi$ are destruction and creation signal bose -operators and $c, c^+ : \Phi \to \Phi$ are these of an external radiation interacting with an active nonlinear media of a device under consideration. The expression (18.1) is a simplification of the following two-modes quantum Hamiltonian

$$\hat{H} = \sum_{j=\overline{1,2}} \hbar\omega_j a_j^+ a_j + \hbar\omega_0 c^+ c - \hbar\chi_0 a_1^+ a_2^+ a_2 a_1 + \chi(a_1^+ a_2^+ c + c^+ a_1 a_2)\,,$$

where a_1-operators specify "signal" states, a_2-operators specify "idler' states in the crystal, describing a nonlinear down conversion optical amplifier, stabilized via the Kerr effect interaction.

If the coupling parameter in (18.1) is taken as $\chi = 0$, then the nonperturbed Hamiltonian

$$\hat{H}_0 = \hbar\omega_0 a^+ a - \chi_0 a^+ a^2 a^+ + \hbar\omega_1 c^+ c \tag{18.2}$$

is not quadratic in (a, a^+)-operator terms, but is such in the $(a^+ a)$-terms:

$$\hat{H}_0 = \hbar(\omega_0 - \chi_0) a^+ a - \hbar\chi_0 (a^+ a)^2 + \hbar\omega_1 c^+ c \tag{18.3}$$

The latter obviously suggests a way for finding its spectrum in exact form based on that of the particle number operators $\hat{N}_0 = a^+ a$, $\hat{N}_1 = c^+ c$:

$$\hat{H}_0 |\Omega_{m,n}\ \rangle = E_{m,n} |\Omega_{m,n}\ \rangle\,, \tag{18.4}$$

where $|\Omega_{m,n}\ \rangle \in \Phi$, $m, n \in \mathbb{Z}_+$, and

$$E_{m,n} = \hbar(\omega_0 - \chi_0) m - \hbar\chi_0 m^2 + \hbar\omega_1 n \tag{18.5}$$

as $a^+ a |\Omega_{m,n}\ \rangle = m |\Omega_{m,n}\ \rangle$ and $c^+ c |\Omega_{m,n}\ \rangle = n |\Omega_{m,n}\ \rangle$ for all $m, n \in \mathbb{Z}_+$, can easily be constructed as follows:

$$|\Omega_{m,n}\ \rangle = \frac{(c^+)^n (a^+)^m}{\sqrt{m!}\sqrt{n!}} |0\ \rangle\,, \tag{18.6}$$

where $|0\ \rangle \in \Phi$ is the standard vacuum state of the Fock space Φ. Hence, we are now in a position to describe the ground state $|\Omega\rangle \in \Phi$ of our dynamical system (18.1) under nonzero interaction. Since the Hamiltonian operator describes a coherent absorption of an external radiation of frequency ω_1 and formation of signal photon states of frequency ω_0 the total radiative energy is not conserved, meaning therefore that the process described by the Hamiltonian (18.1) should be considered as dissipative.

Nevertheless, the task of finding the equilibrium ground state of the model Hamiltonian (18.1) is still in force owing to the non-conservation of both signal and radiation photons generated by laser. Thus, we can consider our Hamiltonian (18.1) in the chemical potential expanded form

$$\hat{H}_\mu = \hbar\omega_0 a^+ a - \chi_0 a^+ a^2 a^+ + \chi(a^+ c + c^+ a) + \hbar\omega_1 c^+ c - \mu_0 a^+ a - \mu_1 c^+ c\,, \tag{18.7}$$

where $\mu_0, \mu_1 \in \mathbb{R}^1$ is a chemical potential of our system at the equilibrium ground state $|\Omega_\mu\rangle \in \Phi$. It can be determined from the following characteristic equilibrium conditions imposed upon the system:

$$-\frac{\partial E_\mu}{\partial \mu_0} = \langle\, \Omega_\mu | a^+ a | \Omega_\mu \,\rangle = N_0\,,$$
$$-\frac{\partial E_\mu}{\partial \mu_1} = \langle\, \Omega_\mu | c^+ c | \Omega_\mu \,\rangle = N_1\,, \tag{18.8}$$

with $N_0 \in \mathbb{Z}_+$ being the equilibrium number of activated with signal photons molecular states in a crystal and $N_1 \in \mathbb{Z}_+$ is the equilibrium number of external laser radiated photon states inside a crystal due to interaction. One can now observe (from (18.8)) that the total number

$$N_0 + N_1 = \bar{N}_1\,, \tag{18.9}$$

is constant, and equals the total number of radiated photons by laser. This equality is clearly true only in the case when the total amount $\bar{N}_0$ of active molecular states inside a crystal is greater than that amount $\bar{N}_1$ of photons radiated by laser. Anyway, the inequality $\bar{N}_1 \leq \bar{N}_0$ is enjoyed in general by most of quantum-optical electronics devices. The condition (18.9) in the form

$$-\left(\frac{\partial E_\mu}{\partial \mu_0}\right) + \left(\frac{\partial E_\mu}{\partial \mu_1}\right) = \bar{N}_1 \tag{18.10}$$

should be augmented still by another important physical condition

$$\left(\frac{\partial E_\mu}{\partial N_0}\right) = 0\,, \tag{18.11}$$

meaning obviously, that the ground state $|\Omega_\mu\,\rangle \in \Phi$ must be specified by the least energy $E_\mu \in \mathbb{R}$ subject to the amount of emitted signal photons $N_0 \in \mathbb{Z}_+$ owing to the interaction of a crystal with laser radiation. Two conditions (18.10) and (18.11) constitute a complete set of constraints imposed upon our ground state $|\Omega_\mu\,\rangle \in \Phi$ to be determined uniquely. For the above constraints to be implemented analytically we need to develop a technique for finding the quantities (18.8) in the functional form suitable for analysis. This can be done based on the Feynman diagrammatic approach to interacting quantum fields.

Appendix A

The Green's Function

As is well known, the most effective method of finding the ground state characteristics of the Hamiltonian (18.1) under constraints (18.10) and (18.11) is making use of Feynman diagrammatic approach subject to the interaction Hamiltonian $\hat{H}_{int} = \hat{H}_\mu - \hat{H}_{0,\mu}$ in the interaction representation. Especially it proved to have been useful for calculating the standard Green's functions in the Heisenberg representation:

$$\begin{aligned} iG_0(t,t') &= \langle\, \Omega_\mu | T(a_{H_\mu}(t) a^+_{H_\mu}(t')) | \Omega_\mu \,\rangle\,, \\ iG_1(t,t') &= \langle\, \Omega_\mu | T(c_{H_\mu}(t) c^+_{H_\mu}(t')) | \Omega_\mu \,\rangle\,, \end{aligned} \tag{A.1}$$

and so-called the anomalous Green's function

$$iG_{01}(t,t') = \langle\, \Omega_\mu | T(c_{H_\mu}(t) a^+_{H_\mu}(t')) | \Omega_\mu \,\rangle\,, \tag{A.2}$$

where the operation "T" denotes the standard chronological operator ordering, giving rise to the following expression for the ground state energy $E_\mu \in \mathbb{R}$:

$$E_\mu = \langle\, \Omega_\mu | \hat{H}_\mu | \Omega_\mu \,\rangle = \langle\, \Omega_\mu | \hat{H}_{0,\mu} | \Omega_\mu \,\rangle - i\chi (G_{01}(t,t^+) + G^*_{01}(t,t^+))|_{t=0}\,. \tag{A.3}$$

Owing to the following simply verified general expression

$$E_\mu = E_{0,\mu} + \int_0^1 d\ln\lambda \langle\, \Omega_\mu(\lambda) | \lambda \hat{H}_{int} | \Omega_\mu(\lambda) \,\rangle\,, \tag{A.4}$$

where $|\Omega_\mu(\lambda)\,\rangle \in \Phi$ is the ground state corresponding to the scaled interaction Hamiltonian $\lambda \hat{H}_{int} : \Phi \to \Phi$, and $E_{0,\mu} \in \mathbb{R}$ is the ground state energy of the nonperturbed Hamiltonian $\hat{H}_{0,\mu} : \Phi \to \Phi$, one can determine (A.3) only due to the (A.2) redefined as

$$iG_{01,\lambda}(t,t') = \langle\ \Omega_\mu(\lambda)|T(c_{H_\mu(\lambda)}(t)a^+_{H_\mu(\lambda)}(t'))|\Omega_\mu(\lambda)\ \rangle\,, \tag{A.5}$$

where $\lambda \in (0,1]$ is a scaling parameter. Namely, from (A.4) and (A.5) one follows that

$$E_\mu = E_{0,\mu} - i\chi \int_0^1 d\lambda (G_{01,\lambda}(t,t^+) + G^*_{01,\lambda}(t,t^+))|_{t=0}\,, \tag{A.6}$$

where the energy $E_{0,\mu}$ is found easily owing to the expression (18.6):

$$E_{0,\mu} = \hbar(\omega_0 - \chi_0 - \mu_0/\hbar)N_0 - \hbar\chi_0 N_0^2 + \hbar\omega_1(N_1 - \mu_1/\hbar)\,. \tag{A.7}$$

Whence we arrived at the need to calculate just the anomalous Green's function (A.5) subject to the scaled interaction Hamiltonian $\lambda\hat{H}_{int} : \Phi \to \Phi$ for all $\lambda \in (0,1]$. We shall next discuss this problem in detail.

Appendix B

The Feynman Diagram Approach: Perturbation Series Expansion

Subject to the Green's functions (A.1) and (a.2) we need to define the nonperturbed ground state $|\Omega_{0,\mu}\rangle \in \Phi$, enjoying the equation

$$\hat{H}_{0,\mu}|\Omega_{0,\mu}\rangle = E_{0,\mu}|\Omega_{0,\mu}\rangle, \tag{B.1}$$

where $E_{0,\mu}$ is given by the expression (A.7). Owing to the fact that photons are bose-particles, we are forced to make shifts in (a, a^+) and (c, c^+)-operators producing so called Bose condensates subject to constraints (18.8):

$$\begin{aligned} a :&\Longrightarrow \tilde{a} + \sqrt{N_0},\ a^+ :\Longrightarrow \tilde{a}^+ + \sqrt{N_0}, \\ c :&\Longrightarrow \tilde{c} + \sqrt{N_1},\ c^+ :\Longrightarrow \tilde{c}^+ + \sqrt{N_1}, \end{aligned} \tag{B.2}$$

where operators $\tilde{a}$ and $\tilde{c}$ enjoy the following no-particle vacuum state conditions:

$$\tilde{a}|\Omega_{0,\mu}\rangle = 0,\ \tilde{c}|\Omega_{0,\mu}\rangle = 0. \tag{B.3}$$

Subject to the transformations (B.2) our interaction Hamiltonian $\hat{H}_{int} : \Phi \to \Phi$ becomes

$$\begin{aligned} \hat{H}_{int} &= \chi[(\tilde{a}^+ + \sqrt{N_0})(\tilde{c} + \sqrt{N_1}) + (\tilde{a} + \sqrt{N_0})(\tilde{c}^+ + \sqrt{N_1})] \\ &= \chi[(\tilde{a}^+\tilde{c} + \tilde{a}\tilde{c}^+) + (\tilde{c}\sqrt{N_0} + \tilde{a}\sqrt{N_1}) \\ &\quad + (\tilde{c}^+\sqrt{N_0} + \tilde{a}^+\sqrt{N_1})] + 2\chi\sqrt{N_0 N_1}. \end{aligned} \tag{B.4}$$

This means that the nonperturbed Hamiltonian $\hat{H}_{0,\mu}$ should be modified by the constant term $2\chi\sqrt{N_0 N_1}$ giving rise to the expression:

$$\begin{aligned} \tilde{H}_{0,\mu} &= \hbar(\omega_0 - \mu_0/\hbar)(\tilde{a}^+ + \sqrt{N_0})(\tilde{a} + \sqrt{N_0}) + \hbar(\omega_1 - \mu_1/\hbar) \times \\ &\quad \times (\tilde{c}^+ + \sqrt{N_1})(\tilde{c} + \sqrt{N_1}) + \chi_0(\tilde{a}^+ + \sqrt{N_0}) \\ &\quad \times (\tilde{a} + \sqrt{N_0})^2(\tilde{a}^+ + \sqrt{N_0}) + 2\chi\sqrt{N_0 N_1}. \end{aligned} \tag{B.5}$$

Thus the scaled modified interaction Hamiltonian becomes

$$\lambda\tilde{H}_{int} = \lambda\chi[(\tilde{a}^+\ \tilde{c} + \tilde{a}\ \tilde{c}^+) + (\tilde{c} +\ \tilde{c}^+)\sqrt{N_0} + (\tilde{a}^+ + \tilde{a})\sqrt{N_1}]\,, \tag{B.6}$$

where $\lambda \in (0, 1]$.

Consider now the modified ground state $|\tilde{\Omega}_\mu\ \rangle \in \Phi$, satisfying the equation like (B.1):

$$\tilde{H}_{0,\mu}|\tilde{\Omega}_{0,\mu}\ \rangle = \tilde{E}_{0,\mu}|\tilde{\Omega}_{0,\mu}\ \rangle\,. \tag{B.7}$$

Since the transformation (B.2) is linear the spectrum of the operator (B.5) coincides with that of the operator $\hat{H}_{0,\mu}$ up to the constant shift $2\chi\sqrt{N_0 N_1}$. Thus,

$$\begin{aligned}\tilde{E}_{0,\mu} &= (\hbar\omega_0 - \chi_0 - \mu_0/\hbar)N_0 - \hbar\chi_0 N_0^2 \\ &\quad + \hbar\omega_1(N_1 - \mu_1/\hbar) + 2\chi\sqrt{N_0 N_1}\,,\end{aligned} \tag{B.8}$$

and we arrive at the starting point of the formula (A.4):

$$\tilde{E}_\mu = \tilde{E}_{0,\mu} + \int_0^1 d\ln\lambda \langle\ \tilde{\Omega}_\mu(\lambda)|\lambda\tilde{H}_{int}|\tilde{\Omega}_\mu(\lambda)\ \rangle\,, \tag{B.9}$$

where the term $\lambda\tilde{H}_{int}$ is given by (B.6). To calculate the matrix element in (B.9) analytically, it is necessary to find all Green's functions modifying (A.1) and (A.2):

$$\begin{aligned}i\tilde{G}_{0,\lambda}(t,t') &= \langle\ \tilde{\Omega}_\mu(\lambda)|T(\tilde{a}_{\tilde{H}_\mu}(t)\tilde{a}^+_{\tilde{H}_\mu}(t'))|\tilde{\Omega}_\mu(\lambda)\ \rangle\,, \\ i\tilde{G}_{1,\lambda}(t,t') &= \langle\ \tilde{\Omega}_\mu(\lambda)|T(\tilde{c}_{\tilde{H}_\mu}(t)\tilde{c}^+_{\tilde{H}_\mu}(t'))|\tilde{\Omega}_\mu(\lambda)\ \rangle\,,\end{aligned} \tag{B.10}$$

and

$$i\tilde{G}_{01,\lambda}(t,t') = \langle\ \tilde{\Omega}_\mu(\lambda)|T(\tilde{a}^+(t')\tilde{c}(t))|\tilde{\Omega}_\mu(\lambda)\ \rangle\,. \tag{B.11}$$

Owing to the general Feynman rules of finding Green's functions (B.10) and (B.11) it is necessary to calculate first all nonperturbed Green's functions at $\chi = 0$; that is, functions

$$\begin{aligned}i\tilde{G}_0^{(0)}(t,t') &= \langle\ \tilde{\Omega}_{0,\mu}|T(\tilde{a}_{\tilde{H}_{0,\mu}}(t)\tilde{a}^+_{\tilde{H}_{0,\mu}}(t'))|\tilde{\Omega}_{0,\mu}\ \rangle\,, \\ i\tilde{G}_1^{(0)}(t,t') &= \langle\ \tilde{\Omega}_{0,\mu}|T(\tilde{c}_{\tilde{H}_{0,\mu}}(t)\tilde{c}^+_{\tilde{H}_{0,\mu}}(t'))|\tilde{\Omega}_{0,\mu}\ \rangle\end{aligned} \tag{B.12}$$

and

$$i\tilde{G}_{01}^{(0)}(t,t') = \langle\ \tilde{\Omega}_{0,\mu}|T(\tilde{c}_{\tilde{H}_{0,\mu}}(t)\tilde{a}^+_{\tilde{H}_{0,\mu}}(t'))|\tilde{\Omega}_{0,\mu}\ \rangle \tag{B.13}$$

for all times $t, t' \in \mathbb{R}^1$. Due to the result (B.8) the Green's functions can be found explicitly after some simple (but a bit cumbersome) calculations as follows: for photon creation-annihilation Green's function one gets

$$\begin{aligned}\tilde{G}_0^{(0)}(t,t') &= N_0 \exp[i(t-t^{'})(\omega_0 - \chi_0 - \mu_0/\hbar - 2\chi_0 N_0/\hbar)] \\ &\quad + \vartheta(t-t^{'}) + N_0\{\exp[it(\omega_0 - \chi_0 - \mu_0/\hbar - 2\chi_0 N_0/\hbar)] - 1\} \\ &\quad \times \{\exp[-it'(\omega_0 - \chi_0 - \mu_0/\hbar - 2\chi_0 N_0/\hbar)] - 1\}, \qquad \text{(B.14)}\end{aligned}$$

$$\begin{aligned}\tilde{G}_1^{(0)}(t,t') &= N_1 \exp[i(t-t^{'})(\omega_1 - \mu_1/\hbar)]\vartheta(t-t^{'}) \\ &\quad + N_1\{\exp[it(\omega_1 - \mu_1/\hbar)] - 1\}\{\exp[-it^{'}(\omega_1 - \mu_1/\hbar)] - 1\}, \qquad \text{(B.15)}\end{aligned}$$

and for the anomalous Green's function one gets

$$\tilde{G}_{01}^{(0)}(t,t') = 0\,. \qquad \text{(B.16)}$$

Thereby we can now express Green's functions (A.1) via Feynman diagram series making use of the Dyson equations:

$$\begin{aligned}\tilde{G}_{0,\lambda}(t,t') &= \tilde{G}_0^{(0)}(t,t') + \int\limits_{\mathbb{R}} d\tau^{'} \int\limits_{\mathbb{R}} d\tau^{"} \tilde{G}_0^{(0)}(t,\tau) \widetilde{\sum}^{*}_{0,\lambda}(\tau,\tau^{'})\tilde{G}_{0,\lambda}(\tau^{'},t')\,, \\ \tilde{G}_{1,\lambda}(t,t') &= \tilde{G}_1^{(0)}(t,t') + \int\limits_{\mathbb{R}} d\tau^{'} \int\limits_{\mathbb{R}} d\tau^{"} \tilde{G}_0^{(0)}(t,\tau) \widetilde{\sum}^{*}_{1,\lambda}(\tau,\tau^{'})\tilde{G}_{1,\lambda}(\tau^{'},t')\,, \qquad \text{(B.17)}\end{aligned}$$

with kernels $\widetilde{\sum}^{*}_{0,\lambda}$ and $\widetilde{\sum}^{*}_{1,\lambda}$ called proper self-energy insertions (or mass operators) and expressed in the following diagram form:

$$-i\widetilde{\sum}^{*}_{0,\lambda}(\tau,\tau')$$

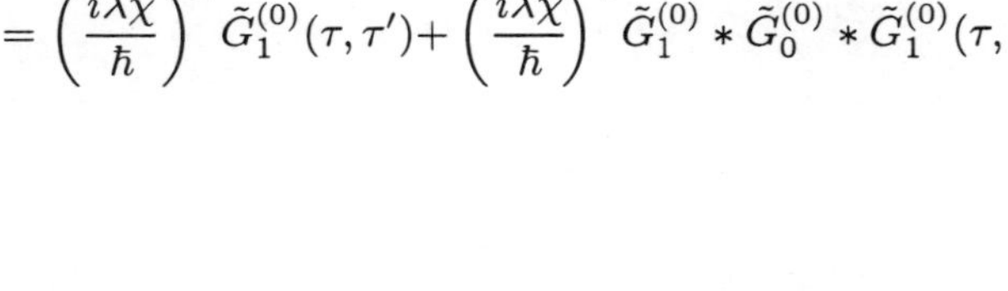

$$= \left(\frac{i\lambda\chi}{\hbar}\right)^2 \tilde{G}_1^{(0)}(\tau,\tau') + \left(\frac{i\lambda\chi}{\hbar}\right)^4 \tilde{G}_1^{(0)} * \tilde{G}_0^{(0)} * \tilde{G}_1^{(0)}(\tau,\tau') + \cdots,$$

$$-i\widetilde{\sum_{1,\lambda}}^{*}(\tau,\tau')$$

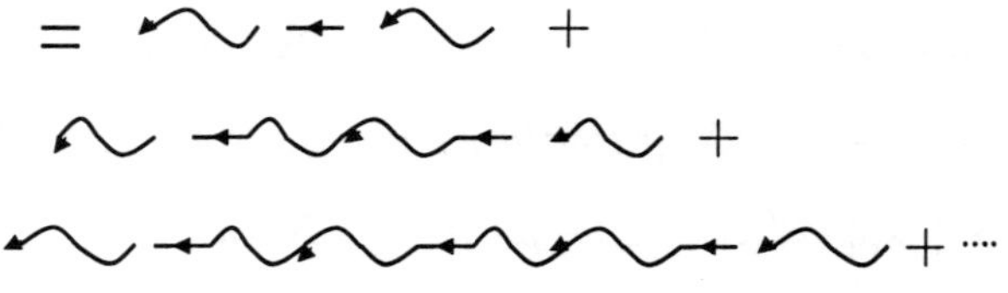

$$= \left(\frac{i\lambda\chi}{\hbar}\right)^2 \tilde{G}_0^{(0)}(\tau,\tau') + \left(\frac{i\lambda\chi}{\hbar}\right)^4 \tilde{G}_0^{(0)} * \tilde{G}_1^{(0)} * \tilde{G}_0^{(0)}(\tau,\tau') + \cdots\,, \tag{B.18}$$

where we denoted by "*" the convolution operation between Green's functions, and

$$\tilde{G}_0^{(0)}(t,t') =$$  , $$\tilde{G}_1^{(0)}(t,t') =$$  . (B.19)

Thereby we can now use of the constraints (18.10) and (18.11) for determining chemical potentials μ_0 and μ_1. Based on expressions (B.17) and (B.18) one arrives at the following integral relationships for Green's functions (B.10):

$$\tilde{G}_{0,\lambda}(t,t') = \tilde{G}_0^{(0)}(t,t') + \left(\frac{\lambda\chi}{\hbar}\right)^2 \tilde{G}_0^{(0)} * \widetilde{\mathcal{R}}^*_{0,\lambda}(t,t')\,,$$

$$\widetilde{\mathcal{R}}^*_{0,\lambda} = \left(\frac{\lambda\chi}{\hbar}\right)^2 \tilde{G}_1^{(0)} * \tilde{G}_0^{(0)} + \left(\frac{\lambda\chi}{\hbar}\right)^2 \tilde{G}_1^{(0)} * \tilde{G}_0^{(0)} * \widetilde{\mathcal{R}}^*_{0,\lambda}\,; \tag{B.20}$$

$$\tilde{G}_{1,\lambda}(t,t') = \tilde{G}_1^{(0)}(t,t') + \left(\frac{\lambda\chi}{\hbar}\right)^2 \tilde{G}_1^{(0)} * \widetilde{\mathcal{R}}^*_{1,\lambda}(t,t')\,,$$

$$\widetilde{\mathcal{R}}^*_{1,\lambda} = \left(\frac{\lambda\chi}{\hbar}\right)^2 \tilde{G}_0^{(0)} * \tilde{G}_1^{(0)} + \left(\frac{\lambda\chi}{\hbar}\right)^2 \tilde{G}_0^{(0)} * \tilde{G}_1^{(0)} * \widetilde{\mathcal{R}}^*_{1,\lambda}\,. \tag{B.21}$$

Similar expressions to the above can be written down for the Green's function (B.13) as well. But we shall not dwell that here.

Appendix C

The Equilibrium State Analysis

As was mentioned before, to specify the equilibrium state $|\tilde{\Omega}_\mu\ \rangle \in \Phi$ completely, we need to implement constraints (18.10) and (18.11) on the energy value $\tilde{E}_\mu$ subject to chemical potentials μ_0 and μ_1. Since the nonperturbed part $\tilde{E}_{0,\mu}$ (B.8) of this energy is already determined, we need calculate only the integral part of (B.9):

$$\Delta\tilde{E}_\mu = \int_0^1 d\ln\lambda\langle\ \tilde{\Omega}_\mu(\lambda)|\lambda\tilde{H}_{int}|\tilde{\Omega}_\mu(\lambda)\ \rangle$$

$$= \chi\int_0^1 d\lambda\langle\ \tilde{\Omega}_\mu(\lambda)|[(\tilde{a}^+\ \tilde{c} + \tilde{a}\ \tilde{c}^+)$$

$$+\sqrt{N_1}(\tilde{a}^+ + \tilde{a}) + \sqrt{N_0}(\tilde{c}^+ + \tilde{c})]|\tilde{\Omega}_\mu(\lambda)\ \rangle$$

$$= i\chi\int_0^1 d\lambda[\tilde{G}_{01,\lambda}(0,0^+) + \tilde{G}^*_{01,\lambda}(0,0^+)] + \chi\sqrt{N_1}\int_0^1 d\lambda[\tilde{A}^*_\lambda(0) + \tilde{A}_\lambda(0)]$$

$$+\chi\sqrt{N_0}\int_0^1 d\lambda[\tilde{C}^*_\lambda(0) + \tilde{C}_\lambda(0)]\,, \tag{C.1}$$

where we defined for any $t \in \mathbb{R}^1$

$$\begin{aligned} \tilde{A}_\lambda(t) &= \langle\ \tilde{\Omega}_\mu(\lambda)|\tilde{a}^+_{\tilde{H}_\mu(\lambda)}(t)|\tilde{\Omega}_\mu(\lambda)\ \rangle\,, \\ \tilde{C}_\lambda(t) &= \langle\ \tilde{\Omega}_\mu(\lambda)|\tilde{c}^+_{\tilde{H}_\mu(\lambda)}(t)|\tilde{\Omega}_\mu(\lambda)\ \rangle. \end{aligned} \tag{C.2}$$

The anomalous functions (C.2) can be easily calculated making use of the Feynman diagram expansions:

$\tilde{A}_\lambda(t) =$

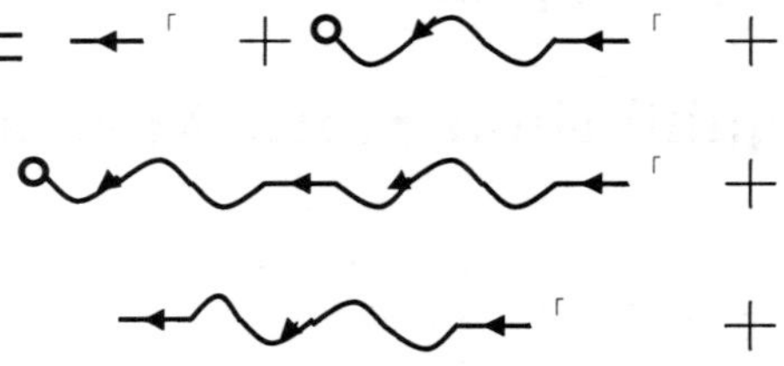

$$\Longrightarrow \sqrt{N_1}\left(\frac{i\lambda\chi}{\hbar}\right)\tilde{G}_0^{(0)}(\cdot,t) + \sqrt{N_1}\left(\frac{i\lambda\chi}{\hbar}\right)^3\tilde{G}_0^{(0)}(\cdot) *$$

$$*\tilde{G}_1^{(0)} * \tilde{G}_0^{(0)}(\cdot,t) + \cdots + \sqrt{N_0}\left(\frac{i\lambda\chi}{\hbar}\right)^2\tilde{G}_1^{(0)}(\cdot) * \tilde{G}_0^{(0)}(\cdot,t)$$

$$+\sqrt{N_0}(\frac{i\lambda\chi}{\hbar})^4\tilde{G}_1^{(0)}(\cdot) * \tilde{G}_0^{(0)} * \tilde{G}_1^{(0)} * \tilde{G}_0^{(0)}(\cdot,t) + \cdots$$

$$= \tilde{A}_{0,\lambda}(t) + \tilde{A}_{1,\lambda}(t)\,,$$

$\tilde{C}_\lambda(t) =$

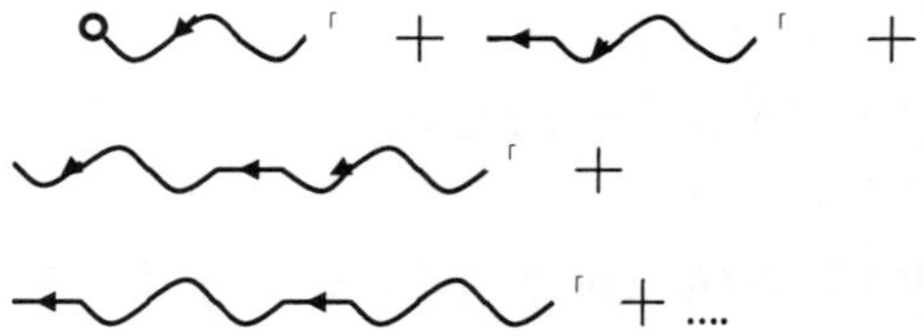

$$\Longrightarrow \sqrt{N_0}\left(\frac{i\lambda\chi}{\hbar}\right)\tilde{G}_1^{(0)}(\cdot,t) + \sqrt{N_1}\left(\frac{i\lambda\chi}{\hbar}\right)^2\tilde{G}_0^{(0)}(\cdot) * \tilde{G}_1^{(0)}(\cdot,t) + \cdots$$

$$+\sqrt{N_0}\left(\frac{i\lambda\chi}{\hbar}\right)^3\tilde{G}_1^{(0)}(\cdot) * \tilde{G}_0^{(0)} * \tilde{G}_1^{(0)}(\cdot,t) + \cdots$$

$$= \tilde{C}_{0,\lambda}(t) + \tilde{C}_{1,\lambda}(t)\,, \qquad \text{(C.3)}$$

where by definition, the sign "⊙" means a factor $\sqrt{N_0}$, the sign "×" means a factor $\sqrt{N_1}$, an internal dot "·" corresponds to a factor $\left(\frac{i\chi}{\hbar}\right)$, with the following Dyson type relationships to be enjoyed:

$$\tilde{A}_{0,\lambda}(t) = \sqrt{N_1}\left(\frac{i\lambda\chi}{\hbar}\right)\tilde{G}_0^{(0)}(\cdot,t) + \left(\frac{i\lambda\chi}{\hbar}\right)^2 \tilde{A}_{0,\lambda}(t) * \tilde{G}_1^{(0)} * \tilde{G}_0^{(0)}(\cdot,t)\,,$$

$$\tilde{A}_{1,\lambda}(t) = \sqrt{N_0}\left(\frac{i\lambda\chi}{\hbar}\right)^2 \tilde{G}_1^{(0)}(\cdot) * \tilde{G}_0^{(0)}(\cdot,t) + \left(\frac{i\lambda\chi}{\hbar}\right)^2 \tilde{A}_{1,\lambda}(t) * \tilde{G}_1^{(0)} * \tilde{G}_0^{(0)}(\cdot,t)\,;$$

$$\tilde{C}_{0,\lambda}(t) = \sqrt{N_1}\left(\frac{i\lambda\chi}{\hbar}\right)^2 \tilde{G}_0^{(0)}(\cdot) * \tilde{G}_1^{(0)}(\cdot,t) + \left(\frac{i\lambda\chi}{\hbar}\right)^2 \tilde{C}_{0,\lambda} * \tilde{G}_0^{(0)} * \tilde{G}_1^{(0)}(\cdot,t)\,,$$

$$\tilde{C}_{1,\lambda}(t) = \sqrt{N_0}\left(\frac{i\lambda\chi}{\hbar}\right)\tilde{G}_1^{(0)}(\cdot,t) + \left(\frac{i\lambda\chi}{\hbar}\right)^2 \tilde{C}_{1,\lambda} * \tilde{G}_0^{(0)} * \tilde{G}_1^{(0)}(\cdot,t)\,. \tag{C.4}$$

Having solved the equations (C.3) and (C.4) we arrive (due to expressions (C.2) and (C.1)) at the energy shift $\Delta\tilde{E}_\mu$ which makes it possible to satisfy constraints (18.10) and (18.11), where E_μ should be replaced by $\tilde{E}_\mu$:

$$E_\mu \Longrightarrow \tilde{E}_\mu = \tilde{E}_{0,\mu} + \Delta\tilde{E}_\mu \tag{C.5}$$

and $\tilde{E}_{0,\mu}$ is given by the expression (B.8). In the case when the susceptibility parameter $\chi \in \mathbb{R}_+$ is too small, the equations (C.3) and (C.4) can easily be solved, where by making it possible to introduce the constraints (18.10) and (18.11) into the analysis of the ground state energy $E_\mu \in \mathbb{R}$. This can be done self-consistently by means of simple calculations which we shall not discuss here.

Bibliography

Abrikosov A., Gorkov L., Dzyaloshinsky I. Methods of quantum field theory in statistical physics. Dover Publ. Inc., NY (1963)

Allen L., Eberly J. Optical Resonance and Two-level Atoms. Oxfard Publ. (1974)

Bogoliubov N.N., Bogoliubov N.N. (jr) Introduction to quantum statistical mechanics, World Scientific (1986)

Dirac P.A.M. The principles of quantum mechanics. Oxford University Press, Oxford (1947)

Fetter A., Walecka J. Quantum theory of many-particle systems. Mc Graw Hill, NY (1971)

Feynman R., Hibbs A. Quantum mechanics and path integrals. Mc Graw Hill, NY (1965)

Gell-Mann and Brueckner, Phys.Rev. 106 p. 364, (1957)

Kac M. Probability and related topics in physics sciences. Interscience, NY (1959)

Kleinert H. Path integrals in quantum mechanics, statistics and polymer physics. World Scienetific (1995)

Prykarpatsky A., Mykytiuk I. Algebraic integrability of nonlinear dynamical systems on manifolds: classical and quantum aspects. Kluwer Publ., the Netherlands (1998)

Reed M., Simon B. Methods of modern mathematical physics, Springer (1974)

Rupasov V.P On theory of the Dicke Super-Radiance.The exact solution of Quasi-One-Dimensional Quantum Model. Journal of Experimental and Theoretical Physics.V.83, N5, p1711-1723 (1982)

Scully M.O., Zubairy M.S. Quantum optics. Cambridge University Press, Cambridge (1986)

Shumovsky A., Yukalov V. Lectures on phase transitions. World Scientific (1990)

Taneri U., Huseyin K., Yu P. Analysis of bifurcations and stability properties of molecular systems. Dynamics and Stability of systems, v.10, N2 p.145-161. (1995)

Walls D.F., Milburn G.J. Quantum optics. Springer-Verlag (1984)

Index